More Praise for *Out of Gas*

"If you want to get a look at the whole forest at once, Goodstein's little book will float you appropriately high above the trees." —*American Scientist*

"Goodstein's book is not a happy read, but an important one. In layman's terms, he explains the science behind his prediction and why other fossil fuels might not do the trick." —*Newsweek*

"*Out of Gas* . . . isn't merely a polemic. Rather, it's a well-reasoned argument that the world needs to quickly come up with new ideas about energy. . . . A book that Dick Cheney's energy task force should have read, and it should be mailed to Gov. Rick Perry's Energy Planning Council. It's an eye-opening look at what could be the world's single biggest challenge." —Bill Day, *San Antonio Express-News*

"A short, fascinating work meant as a primer on 'the relevant laws of nature.' " —Eli Sanders, *Seattle Times*

W. W. Norton & Company New York London

DAVID GOODSTEIN

OUT OF GAS

THE END OF THE
AGE OF OIL

Chapter 1 will also be published, in slightly different form, as "Energy, Technology and Climate: Running out of Gas," in Galston, Arthur and Christiana Peppard, eds., *Expanding Horizons in Bioethics* (Dordrecht: Kluwer Academic Publishers, 2005).

For information about permission to reproduce selections from this book, write to Permissions, W. W. Norton & Company, Inc., 500 Fifth Avenue, New York, NY 10110

Manufacturing by The Courier Companies, Inc.
Book design by Nan Oshin
Photography and Illustration by Jeffrey Atherton
Production manager: Andrew Marasia

Library of Congress Cataloging-in-Publication Data
Goodstein, David L., 1939–
    Out of gas : the end of the age of oil / by David Goodstein.
    p. cm.
Includes bibliographical references and index.
    ISBN 0-393-05857-3 (hardcover)
    1. Petroleum reserves. 2. Petroleum industry and trade. 3. Petroleum reserves—Forecasting. 4. Petroleum industry and trade—Forecasting. I. Title.
    TN870.G645 2004
    622'.1828—dc21

                                        2003010376

ISBN 0-393-32647-0 pbk.

W. W. Norton & Company, Inc.
500 Fifth Avenue,
New York, N.Y. 10110
www.wwnorton.com

W. W. Norton & Company Ltd.
Castle House, 75/76 Wells Street, London W1T 3QT

3 4 5 6 7 8 9 0

To our children and grandchildren,
who will not inherit the riches that we inherited

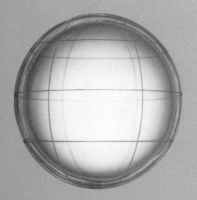

CONTENTS

OUT OF GAS

INTRODUCTION

T he world will soon start to run out of conventionally produced, cheap oil. If we manage somehow to overcome that shock by shifting the burden to coal and natural gas, the two other primary fossil fuels, life may go on more or less as it has been—until we start to run out of all fossil fuels by the end of this century. And by the time we have burned up all that fuel, we may well have rendered the planet unfit for human life. Even if human life does go on, civilization as we know it will not survive, unless we can find a way to live without fossil fuels.

Technically, it might be possible to accomplish that. Power plants can run on nuclear energy or sunlight. Part of that power can be used to generate hydrogen fuel or charge advanced batteries for use in transportation. There are huge technical problems to be solved, certainly, but most of the scientific principles are well understood, and human beings are very good at solving technical problems. In fact if we put our minds to it, we could start trying to kick the fossil fuel habit now, protecting the planet's climate from further damage and preserving the fuels for future generations as the source of chemical goods. Ninety percent of the organic chemicals we use—pharmaceuticals, agricultural chemicals, plastics—are made from petroleum; there are better uses for the stuff than burning it up. To make such an about-face will require global political leadership that is both visionary and courageous. It seems unlikely that we will be so lucky.

Of all the fossil fuels, oil is the most important to us by far. Its existence has been known since ancient times, because of natural seepages at the surface of the earth. Ancient peoples in the Middle East and the Americas used oil for a variety of medicinal, military, and other purposes. It was thought to be useful as a laxative, for example. (Don't try this at home!) And oil-soaked blazing arrows were used by the Persians in the siege of Athens in 480 BCE. But by and large, oil was little needed and little used until the nineteenth century.

By the beginning of the nineteenth century, the growth of urban centers made it necessary to search for better means of illumination. For a while, whale oil lamps served the purpose and whaling became a significant industry. But by the middle of the century, whales were becoming scarce. Kerosene derived from coal was widely used, but a better substitute for whale oil was needed. In August 1859, Edwin L. Drake, a former conductor on the New York & New Haven Railroad, drilled the world's first oil well at a natural seepage near Titusville, in northwestern Pennsylvania. Soon coal-oil refineries were processing cheap oil instead of coal. Then in 1861 the German entrepreneur Nikolaus Otto invented the first gasoline-burning engine (see chapter 4), and soon demand for oil as a fuel began to grow. Within a few decades, oil was being found in and extracted from fields all over the globe. Since E. L. Drake drilled that first well, roughly fifty thousand oil fields have been discovered worldwide.*

In the 1950s, Shell Oil Company geophysicist M. King Hubbert predicted that the rate at which oil could be extracted

---

*Most of those discoveries have been insignificant; about half of all the oil ever discovered was found in the forty largest fields.

from wells in the United States would peak around 1970 and decline rapidly after that. At the time, his prediction was not well received by his peers, but he turned out to be right. U.S. oil extraction peaked at about nine million barrels per day in 1970 and has been declining ever since. Today it's just a little under six million barrels per day. Oil companies now routinely use Hubbert's methods to predict future yields of existing oil fields.

Recently, a number of oil geologists have applied Hubbert's techniques to the oil supply of the entire world. They have each used different data, different assumptions, and somewhat different methods, but their answers have been remarkably similar. The worldwide Hubbert peak, they say, will occur very soon—most probably within this decade.[1] There are highly respected geologists who disagree with that assessment, and the data on which it is based are subject to dispute. Nevertheless, Hubbert's followers have succeeded in making a crucial point: The worldwide supply of oil, as of any mineral resource, will rise from zero to a peak and after that will decline forever.

Some say that the world has enough oil to last for another forty years or more, but that view is almost surely mistaken. The peak, which will occur when we've used half the oil nature made for us (see chapter 1), will come far sooner than that. When the peak occurs, increasing demand will meet decreasing supply, possibly with disastrous results. We had a foretaste of the consequences in 1973, when some Middle Eastern nations took advantage of declining U.S. supplies and created a temporary, artificial shortage. The immediate result was long lines at the gas stations, accompanied by panic and despair for the future of the American way of life. After the worldwide Hubbert's peak, the shortage will not be artificial and it will not be temporary.

There are those who see a silver lining in this dire situation. Since the beginning of the Industrial Revolution, we have been pouring carbon dioxide and other greenhouse gases into the atmosphere precisely because of the burning of fossil fuels. Scientists are fairly certain that the result has been an increase in global temperature, and that this will continue and might accelerate. Could it be that Hubbert's peak will save us from destroying our planet?

The climate of Earth is in a fragile, metastable state that was probably created by life itself. Primitive life forms were responsible for oxygenating the atmosphere, and they were also responsible for laying down huge quantities of carbon, in the form of coal and other fossil fuels. If, after Hubbert's peak, we take to burning coal in large quantities, then Earth's so-called intelligent life will be reconverting that carbon and oxygen into carbon dioxide. We cannot predict exactly what that will do to our climate, but one possibility is that it will throw the planet into an entirely different state. The planet Venus is in such a state: Because of a runaway greenhouse effect, its surface temperature is hot enough to melt lead.

Some economists say that we don't need to worry about running out of oil, because while it's happening the rise in oil prices will make other fuels economically competitive and oil will be replaced by something else. But as we learned in 1973, the effects of an oil shortage can be immediate and drastic, while it may take years, perhaps decades, to replace the vast infrastructure that supports the manufacture, distribution, and consumption of the products of the twenty million barrels of oil we Americans alone gobble up each day.

One certain effect will be steep inflation, because gasoline, along with everything made from petrochemicals and everything that has to be transported, will suddenly cost more. Such an inflationary episode will surely cause severe economic damage—perhaps so severe that we will be unable to replace the world's vast oil infrastructure with something else. That's a prospect we would rather not think about.

Nuclear power plants are so feared and controversial that none have been built in the United States for many years, and some countries (for example, Italy) have outlawed them completely. When the oil crisis comes, opposition to nuclear power is likely to weaken considerably. But it will take at best a decade or more for the first new power plants to come on line—and the use of nuclear fuel is pretty much limited to power plants. Nuclear energy will not easily substitute for oil. Even if we do manage a successful switch to coal or natural gas, we will, in a few reckless generations, have depleted Earth's endowment to us and altered its climate to an extent that we cannot now predict. The only alternative to that dark vision of the future is to learn to live entirely on nuclear power and light as it arrives from the Sun. Will we have the wisdom and ability to do that? If we do, can a civilization as complex as ours live on those resources?

These are big questions, and their answers depend heavily on social and political factors. But there is also a large component of science underlying all of this. We can hope, if we are wise, to alter the laws of peoples. But we cannot change the laws of nature. The intent of this small book is to explain the relevant laws of nature. The idea is to sketch out, for those who are not specialists, both the opportunities and the limitations that nature has provided for us. Only if we understand both can we hope to proceed with wisdom. ●

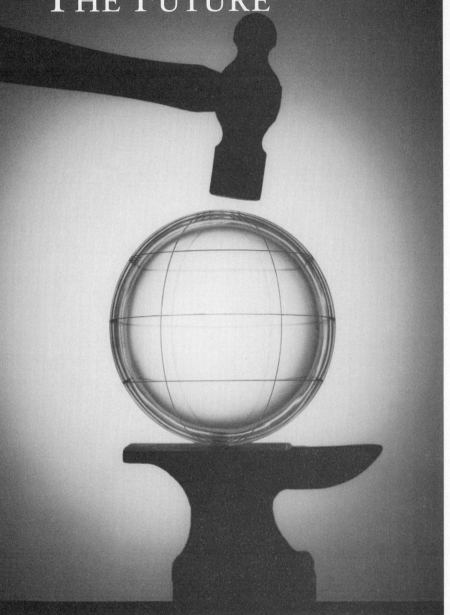
THE FUTURE

I n the 1950s, the United States was the world's leading producer of oil. Much of the nation's industrial and military might derived from its giant oil industry. The country seemed to be floating on a rich, gooey ocean of "black gold." Nobody was willing to believe that the party would ever end. Well, almost nobody. There was a geophysicist named Marion King Hubbert who knew better.

Hubbert, the son of a central Texas farm family, was born in 1903. Somehow he wound up at the far-off University of Chicago, where he earned all his academic degrees right up to the Ph.D. Embittered with the academic world after an unhappy stint teaching geophysics at Columbia University, he spent the bulk of his professional career with the Shell Oil Company of Houston. That's where he was when, in 1956, very much against the will of his employer, he made public his calculation that American oil dominance would soon come to an end.[1] To understand how he reached that conclusion and the relevance of his reasoning to world oil supplies today, we need to understand a bit about how oil came to be in the first place.

For hundreds of millions of years, animal, vegetable, and mineral matter drifted downward through the waters to settle on the floors of ancient seas. In a few privileged places on Earth, strata of porous rock formed that were particularly rich in organic inclusions. With time, these strata were buried deep beneath the seabed. The interior of Earth is hot, heated by the decay of natural radioactive elements. If the porous source rock sank just deep enough, it reached the proper temperature for the organic matter to be transformed into oil. Then the weight of the rock above it could squeeze the oil out of the source rock like water out of a sponge, into layers above and below, where it could be trapped. Over vast stretches of time, in various parts of the globe, the seas retreated, leaving some of those deposits beneath the surface of the land.

Oil consists of long-chain hydrocarbon molecules. If the source rock sank too deep, the excessive heat at greater depths—some three miles below the surface—broke these long molecules into the shorter hydrocarbon molecules we call natural gas. Meanwhile, in certain swampy places, the decay of dead plant matter created peat bogs. In the course of the eons, buried under sediments and heated by Earth's interior, the peat was transformed into coal, a substance that consists mostly of elemental carbon. Coal, oil, and natural gas are the primary fossil fuels. They are energy from the Sun, stored within the earth.

Until only two hundred years ago—the blink of an eye on the scale of our history—the human race was able to live almost entirely on light as it arrived from the Sun. The Sun nourished plants, which provided food and warmth for us and our animals. It illuminated the day and (in most places) left the night sky, sparkling with stars, to comfort us in our repose. Back then, a few people in the civilized world traveled widely, even sailing across the oceans, but most people

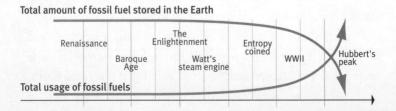

probably never strayed very far from their birthplaces. For the rich, there were beautiful paintings, sophisticated orchestral music, elegant fabrics, and gleaming porcelain. For the common folk, there were more homespun versions of art, music, textiles, and pottery. Merchant sailing ships ventured to sea carrying exotic and expensive cargoes including spices, slaves, and, in summer, ice. At the end of the eighteenth century, no more than a few hundred million people populated the planet. A bit of coal was burned, especially since trees had started becoming scarce in Europe (they soon would begin to disappear in the New World), and small amounts of oil that seeped to the surface found some application, but by and large Earth's legacy of fossil fuels was left untouched.

Today we who live in the developed world expect illumination at night and air conditioning in summer. We may work every day up to a hundred miles from where we live, depending on multiton individual vehicles to transport us back and forth on demand, on roads paved with asphalt (another by-product of the age of oil). Thousands of airline flights per day can take us to virtually any destination on Earth in a matter of hours. When we get there, we can still chat with our friends and family back home, or conduct business as if we had never left the office. Amenities that were once reserved for the rich are available to most people, refrigeration rather than spices preserves food, and machines do much of our hard labor. Ships, planes, trains, and trucks transport goods of every description all around the world. Earth's population exceeds six billion people. We don't see the stars so clearly anymore, but on most counts few of us would choose to return to the eighteenth century.

This revolutionary change in our standard of living did not come about by design. If you asked an eighteenth-century sage like Benjamin Franklin what the world really needed, he would probably not have described those things we have wound up with—except perhaps for the dramatic improvement in public health that has also taken place since then. Instead of design or desire, our present standard of living has resulted from a series of inventions and discoveries that altered our expectations. What we got was not what we wanted or needed but rather what nature and human ingenuity made possible for us. One consequence of those inventions and changed expectations is that we no longer live on light as it arrives from the Sun. Instead we are using up the fuels made from sunlight that Earth stored up for us over those many hundreds of millions of years. Obviously we have unintentionally created a trap for ourselves. We will, so to speak, run out of gas. There is no question about that. There's only a finite amount left in the tank. When will it happen?

Throughout the twentieth century, the demand for and supply of oil grew rapidly. Those two are essentially equal. Oil is always used as fast as it's pumped out of the ground. Until the 1950s, oil geologists entertained the mathematically impossible expectation that the same rate of increase could continue forever. All warnings of finite supplies were hooted down, because new reserves were being discovered faster than consumption was rising. Then, in 1956, Hubbert predicted that the rate at which oil could be extracted from the lower forty-eight United States would peak around 1970 and decline rapidly after that. When his prediction was borne out, other oil geologists started paying serious attention.

Hubbert used a number of methods to do his calculations. The first was similar to ideas that had been used by population biologists for well over a century. When a new population—of humans or any other species—starts growing in an area that has abundant resources, the growth is initially exponential, which means that the rate of growth increases by the same fraction each year, like compound interest in a bank account. That is just how the geologists used to think oil discovery would grow. However, once the population is big enough so that the

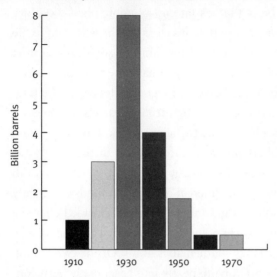

Discovery dates of U.S. oil fields

resources no longer seem unlimited, the rate of growth starts slowing down. The same happens with oil discovery, because the chances of finding new oil get smaller when there's less new oil to find. Hubbert showed that once the rate of increase of oil discovery starts to decline, it's possible to extrapolate the declining rate to find where growth will stop altogether. At that point, all the oil in the ground has been discovered, and the total amount there ever was is equal to the amount that's already been used plus the known reserves still in the ground. Hubbert noticed that the trend of declining annual rate of increase of oil discovery was established for the lower forty-eight states by the 1950s. Others have now pointed out that the rate of discovery worldwide has been declining for decades. The total quantity of conventional oil that Earth stored up for us is estimated by this method to be about two trillion barrels.[2]

Hubbert's second method required assuming that in the long run, when the historical record of the rate at which oil was pumped out of the ground was plotted, it would be a bell-shaped curve. That is, it would first rise (as it has been), then reach a peak that would never be exceeded, and afterward decline forever. Now that it's far enough along, half a century after he made that assumption, it's clear that he was right in the case of the lower forty-eight. If the same assumption is correct for the rest of the world, and if you have the historical record of the rising part of the curve and a good estimate of the total amount of oil that ever was (two trillion barrels, see above), then it's not difficult to predict when the peak, Hubbert's peak, will occur. Hubbert had that information in the 1950s for the lower forty-eight. We have it now for the whole world. Different geologists using different data and methods get slightly different results, but some (not all) have concluded that the peak will happen at some point in this decade. The point can be seen without any fancy mathematics at all. Of the two trillion barrels of oil we started with, nearly half has already been consumed. The peak occurs when we reach the halfway point. That, they say, can't be more than a few years off.

Hubbert's third method applied the observation that the total amount of oil extracted to date paralleled oil discovery but lagged behind by a few decades. In other words, we pump oil out of the ground at about the same rate that we discover it, but a few decades later. Thus the rate of discovery predicts the rate of extraction. Worldwide, remember, the rate of discovery started declining decades ago. In other words, Hubbert's peak for oil discovery already occurred, decades ago. That gives an independent prediction of when Hubbert's peak for oil consumption will occur. It will occur, according to that method, within the next decade or so.

Not all geologists pay attention to this assessment. Many prefer to take the total amount known for sure to be in the ground and divide that by the rate at which it's getting used up. This is known in the industry as the R/P ratio—that is, the ratio of reserves to production. Depending on what data one uses, the R/P ratio is currently between forty and a hundred years. They conclude that if we continue to pump oil out of the ground and consume it at the same rate we are doing now, we will not have pumped the last drop for another forty to one hundred years.

Another point of disagreement concerns the total amount of oil that nature has produced on Earth. Over the period 1995–2000, the United States Geological Survey (USGS) made an exhaustive study of worldwide oil supplies. The resulting report concludes that, with 95 percent certainty, there was the equivalent of at least two trillion barrels when we started pumping. However, it also concludes with 50 percent probability that there were at least 2.7 trillion barrels—based on the expectation that, contrary to trends mentioned earlier, new discovery will continue at a brisk rate for at least thirty more years. The additional 0.7 trillion barrels to be unearthed would amount to discovering all over again all the oil that's now known to exist in the Middle East.

The fact is, the amount of known reserves is a very soft number. For one thing, it is usually a compilation of government or commercial figures from countries around the world, and those reported figures are at least sometimes slanted by political or economic considerations. Also, what we mean by "conventional" or "cheap" oil changes with time. As technology advances, the amount of reserves that can be economically tapped in known fields increases. The way the oil industry uses the term, the increase in recoverable oil counts as "discovery," and it accounts for much of the new discovery the USGS expects in the next thirty years. Finally, as oil starts to become scarce and the price per barrel goes up, the amount recoverable at that price will necessarily also increase.[3] These are all tendencies that might help to push Hubbert's peak farther into the future than the most pessimistic predictions.[4]

Nevertheless, all our experience with the consumption of natural resources suggests that the rate at which we use them up starts at zero, rises to a peak that will never be exceeded, and then declines back to zero as the supply becomes exhausted.

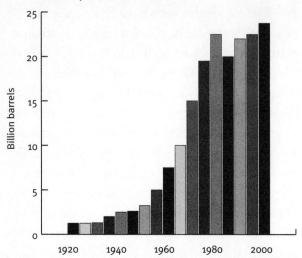

Annual production of world oil

There have been many instances of that behavior: coal mining in Pennsylvania, copper in northern Michigan, and many others, including oil in the lower forty-eight.[5] That picture forms the fundamental basis of the views of Hubbert and his followers, but it is ignored by those who depend on the R/P ratio. Given that worldwide demand will continue to increase, as it has for well over a century, Hubbert's followers expect the crisis to occur when the peak is reached, rather than when the last drop is pumped. In other words, we will be in trouble when we've used up half the oil that existed, not all of it.

If you believe the Hubbert view—that the crisis comes when we reach the production peak rather than the last drop—but you accept the USGS estimate that there may have been an original amount of oil equivalent to 2.7 trillion barrels, then, compared with the earlier estimates, the crisis will be delayed by little more than a decade.* If Hubbert's followers are correct, we may be in for some difficult times in the near future. In an

---

*I have made this calculation using Hubbert techniques. Hubbert represented the rise and fall of oil discovery and extraction by a mathematical form known as the Logistic Curve (also known in business schools as the S-Shaped Curve). Others have used different bell-shaped curves known as Gaussian and Lorentzian Curves. They all give approximately the same results.

orderly, rational world, it might be possible for the gradually increasing gap between supply and demand for oil to be filled by some substitute. But anyone who remembers the oil crisis of 1973 knows that we don't live in such a world, especially when it comes to an irreversible shortage of oil. It's impossible to predict exactly what will happen, but we can all too easily envision a dying civilization, the landscape littered with the rusting hulks of useless SUVs. Worse, desperate attempts by one country or region to maintain its standard of living at the expense of others could lead to Oil War III. (Oil wars I and II are already history.) Knowledge of science is not useful in predicting whether or not such dire events will occur. Science is useful, however, in placing limits on what is possible.

To begin with, conventional oil is not the only oil. Once all the cheap oil is pumped, advanced methods can still squeeze a little more oil out of almost any field. These deposits are known as heavy oil: The more that is extracted, the heavier it gets. There are also deposits of what are known as oil sands and tar sands. Like the remains of depleted oil fields, these deposits are more difficult and expensive to extract; in essence, they are mined and their oil is extracted from the ore rather than just pumped out of the ground.

Then there is shale oil. As noted, conventional oil was created when source rock loaded with organic matter sank just deep enough in the earth for the organics to be cooked properly into oil. Oil shale is source rock that never sank deep enough to make oil. In Colorado, Wyoming, and Utah there is more shale oil than all the conventional oil in the world. However, shale oil in situ is not really oil at all; it was so named in order to attract investment. Instead it is kerogen, a waxy substance that can be made into oil if the rock containing it is mined, crushed, and heated.

Exploiting any of those resources will be more expensive, slower, and more environmentally damaging than pumping conventional oil. It will also require more energy input to get a given amount of energy out. Once the energy needed becomes equal to the energy produced, the game is lost. We are already using in our cars one fuel that may require more energy to produce than it provides: Ethanol made from corn is widely believed to be a net energy loser. As we progress down the fossil fuel list from light crude oil (the stuff we mostly use now) to heavy oil, oil sands, tar sands, and finally shale oil, the cost in energy progressively increases, as do other costs. Some experts believe that shale oil will always be a net energy loser.

Once past Hubbert's peak, as the gap between rising demand and falling supply grows, the rising price of oil may make those alternative fuels economically competitive, but even if they are net energy positive, it may not prove possible to get them into production fast enough to fill the growing gap. That's called the rate-of-conversion problem. Worse, the economic damage done by rapidly rising oil prices may undermine our ability to mount the huge industrial effort needed to get the new fuels into action.

Natural gas, which comes from overcooked source rock, is another alternative in the short term. Natural gas, mostly methane, is relatively easy to extract quickly, and transforma-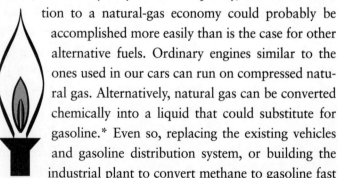tion to a natural-gas economy could probably be accomplished more easily than is the case for other alternative fuels. Ordinary engines similar to the ones used in our cars can run on compressed natural gas. Alternatively, natural gas can be converted chemically into a liquid that could substitute for gasoline.* Even so, replacing the existing vehicles and gasoline distribution system, or building the industrial plant to convert methane to gasoline fast

*Liquid natural gas is a cryogenic (low-temperature) liquid that requires refrigeration and special handling. However natural gas can also be transformed chemically into other fuels that are liquid at normal temperatures.

enough to make up for the missing oil, will be difficult. And even if this transformation is accomplished, success is only temporary. Hubbert's peak for natural gas is estimated to occur only a couple of decades after the one for oil.

There is also a possible fuel called methane hydrate, a solid that looks like ice but burns when ignited. Consisting of methane molecules trapped in a cage of water molecules, it forms when methane combines with water at temperatures close to the freezing point and under high pressure. Methane hydrate was discovered only a few decades ago, and so far there are a number of theories about where it might be found (under the Arctic permafrost, on the deep ocean floor, on the moons of Saturn), how much of it there is, and whether it might be mined (and used) successfully. Not much is known with certainty except that the stuff exists.

A huge amount of chemical potential energy is stored in the earth in the form of elemental carbon, or coal. As is true of the other fossil fuels, to extract the stored energy from coal, each atom of carbon must combine with oxygen to become a molecule of carbon dioxide, a greenhouse gas. But in addition to its inevitable CO_2 production, coal often comes with unpleasant impurities such as sulfur, mercury, and arsenic, none of which can be inexpensively extracted. Coal is a very dirty fuel.* Nevertheless, coal can be liquefied by combining it with hydrogen at high temperature and pressure—an expensive, energy-intensive process that Germany used out of desperation in World War II. If we take our chances on fouling the atmosphere and turn to coal as our primary fuel, we are told that there is enough of it in the ground to last for hundreds of years. That estimate, however, is like the R/P ratio for oil. It doesn't take into account the rising world population, the determination of the developing world to attain a high standard of living,

---

* The dangerous concentrations of mercury that are found in swordfish and tuna originate in coal-fired power plants.

and above all the fact that we now get twice as much energy from oil as from coal and the conversion process is inefficient. Thus if we try to replace the missing oil with liquid fuel from coal, we will have to mine coal on a scale that is almost unimaginable—more than ten times faster than we are doing today. Finally there's the Hubbert's peak effect, which is just as valid for coal as it is for oil. The simple fact is that if we turn to coal as a substitute for oil, the end of the age of fossil fuel, coal included, will probably come in this century.

Controlled nuclear fusion—energy obtained from fusing light nuclei into heavier ones—has long been seen as the ultimate energy source of the future. The technical problems that have prevented successful use of nuclear fusion up to now may someday be solved. Not in time to rescue us from the slide down the other side of Hubbert's peak, certainly, but perhaps someday. Then the fuel—at least initially—would be deuterium, a form of hydrogen found naturally in seawater, and lithium, a light element found in many common minerals.* There would be enough of both to last for a very long time. However, the conquest and practical use of nuclear fusion has proved to be very difficult. It has been said of both nuclear fusion and shale oil that they are the energy sources of the future, and always will be.

Nuclear fission, on the other hand, is a well-established technology. The fuel for this kind of reactor is the highly radioactive isotope uranium-235. (The supply of ^{235}U available for mining will be discussed in chapter 5.) The very word "nuclear" strikes fear into the hearts of many people—so much so that the utterly innocent imaging technique called nuclear magnetic resonance (NMR) by scientists had to be renamed

---

*The nuclear reaction envisioned for fusion reactors is the fusion of deuterium and tritium, two isotopes of hydrogen. Tritium doesn't exist in nature, but the fusion reaction yields neutrons, which would be used to make the tritium in a lithium blanket. Thus the actual fuels are deuterium and lithium. See chapter 5 for further discussion.

magnetic resonance imaging (MRI) before the public could accept it for medical use. When the oil crisis occurs, the fear of nuclear energy is likely to recede, because of the compelling need for it. However, there will continue to be legitimate concerns about safety and nuclear waste disposal. Also, nuclear energy is suitable only for power plants or very large, heavy vehicles, such as ships and submarines. Don't look for nuclear cars or airplanes anytime soon.

What about the possibility that a huge new discovery of conventional oil will put off the problem for the foreseeable future? Better to believe in the tooth fairy. Oil geologists have gone to the ends of the earth searching for oil. There probably isn't enough unexplored territory on this planet to contain a spectacular unknown oil field. The largest remaining accessible and unexplored area is the South China Sea; geologists consider it a promising (but not spectacularly so) region. It is unexplored because of conflicting ownership claims (by China, Taiwan, Vietnam, the Philippines, Malaysia, and Brunei) and murky international law governing mineral rights at sea. Other possible sites that come with big problems include central Siberia and the very deep oceans. Remember that in spite of intense worldwide effort, the rate of oil discovery started declining decades ago and has been in decline ever since. That is why the USGS assumption of thirty more years of rapid discovery seems questionable, even if it is really a prediction about future technology rather than future discovery.

But let us suppose for one euphoric moment that one more really big one is still out there waiting to be discovered. The largest oil field ever found is the Ghawar field in Saudi Arabia, whose eighty-seven billion barrels were discovered in 1948. If someone were to stumble onto another ninety-billion-barrel field tomorrow, Hubbert's peak would be delayed by a year or two, well within the uncertainty of the present estimates of when it will occur. It would hardly make any difference at all. That fact alone points up the sterility of the long-standing debate over drilling for oil in the Arctic National Wildlife Refuge in Alaska. If the ANWR is opened for drilling, its contribution to world oil supplies will be modest indeed. The best reason for not drilling there is not to protect the wildlife. It is to preserve the oil for future generations to use in petrochemicals, rather than burning it up in our cars.

Once Hubbert's peak is reached and oil supplies start to decline, how fast will the gap grow between supply and demand? That is a crucial question, and one that is almost impossible to answer with confidence. Here's a rough attempt at guessing the answer. The upward trend at which the demand for oil has been growing amounts to an increase of a few percent per year. On the other side of the peak, we can guess that the available supply will decline at about the same rate, while the demand continues to grow at that rate. The gap, then, would increase at about, say, 5 percent per year. Therefore, ten years after the peak we would need a substitute for close to half the oil we use today—that is, a substitute for something like ten billion to fifteen billion barrels per year. Even in the absence of any major disruptions caused by the oil shortages after the peak, it is very difficult to see how an effective substitution can possibly be accomplished.

To be sure, the effects of the looming crisis could be greatly mitigated by taking steps to decrease the demand for oil. For example, with little sacrifice of convenience or comfort, we Americans could drive fuel-efficient hybrids rather than humongous gas guzzlers. There are countless other ways in which we could reduce our extravagant consumption of energy: redesigned cities, better insulated homes, improved public transportation, and so on. Such changes are beginning to be made, but there are powerful interests—like the oil companies and the automobile industry and its unions—opposing them.

Before we turn to prospects for the future, a little summing up is in order. The followers of King Hubbert may or may not be correct in their quantitative predictions of when the peak will occur. Regardless of that, they have taught us a very important lesson. The crisis will come not when we pump the last drop of oil but rather when the rate at which oil can be pumped out of the ground starts to diminish. That means the crisis will come when we've used roughly half the oil that nature made for us. Any way you look at it, the problem is much closer than we previously imagined. Moreover, burning fossil fuels alters the atmosphere and could threaten the balmy, metastable state our planet is in. We have some very big problems to solve.

FUTURE SCENARIOS

So, what does the future hold? We can easily sketch out a worst-case scenario and a best-case scenario.

Worst case: After Hubbert's peak, all efforts to produce, distribute, and consume alternative fuels fast enough to fill the gap between falling supplies and rising demand fail. Runaway inflation and worldwide depression leave many billions of people with no alternative but to burn coal in vast quantities for warmth, cooking, and primitive industry. The change in the greenhouse effect that results eventually tips Earth's climate into a new state hostile to life. End of story. In this instance, worst case really means worst case.

Best case: The worldwide disruptions that follow Hubbert's peak serve as a global wake-up call. A methane-based economy is successful in bridging the gap temporarily while nuclear power plants are built and the infrastructure for other alternative fuels is put in place. The world watches anxiously as each new Hubbert's peak estimate for uranium and oil shale makes front-page news.

No matter what else happens, this is the century in which we must learn to live without fossil fuels. Either we will be wise enough to do so before we have to, or we will be forced to do so when the stuff starts to run out. One way to accomplish that would be to return to life as it was lived in the eighteenth century, before we started to use much fossil fuel. That would require, among many other things, eliminating roughly 95 percent of the world's population. The other possibility is to devise a way of running a complex civilization approximating the one we have now which does not use fossil fuel. Do the necessary scientific and technical principles exist?

One of the more difficult problems will be finding a fuel for transportation. One possibility is that advanced electric batteries will make battery-powered electric vehicles practical. In the past decade, batteries packing many times as much energy in a given volume as the batteries commonly used in cars have been developed for use in mobile phones and portable computers. There is no reason why such advanced batteries can't become the basis of future means of transportation.[6] Or the transportation fuel of the future might be hydrogen—not deuterium for thermonuclear fusion but ordinary hydrogen, to be burned as a fuel by old-fashioned combustion or used in hydrogen fuel cells, which produce electricity directly. Burning hydrogen or using it in fuel cells puts into the atmosphere nothing but water vapor.* Water vapor is a greenhouse gas, to be sure, but unlike carbon dioxide it cycles rapidly out of the atmosphere as rain or snow. Hydrogen is dangerous and difficult to handle and store, but so are gasoline and methane. Nature has not set aside a supply for us, but we can make it ourselves.

As the French engineer and thermodynamicist Sadi Carnot understood perfectly (see chapter 4), you can't get something for nothing. Hydrogen is a high-potential-energy substance; that's precisely why it is valuable as a fuel. So is the working fluid of an electric battery. But thermodynamically speaking,

---

*However, hydrogen gas would also inevitably be leaked to the atmosphere. That would pose a threat to Earth's protective ozone layer. Other problems resulting from widespread use of hydrogen will inevitably arise.

hydrogen and batteries are not literally sources of energy. They are only means of storing and transporting it. That energy has to come from somewhere. Where will we get the energy, for example, to make hydrogen? Interestingly, one possible source is the potential energy stored in coal. There are existing processes that combine coal and steam to make hydrogen—and, inevitably, carbon dioxide. Hydrogen gas is produced from a slurry of water and coal using a calcium oxide → calcium carbonate intermediary reaction. The calcium carbonate is then converted back into calcium oxide (to be used over again) and carbon dioxide. In principle, the carbon dioxide could be separated and stored—"sequestered" is the current buzzword. Where could it be sequestered? That problem has not been solved yet (more about this in chapter 5). And in any case the coal will eventually run out, whereas we're trying to think long-term here.

The interior of the earth is heated by the decay of natural radioactive elements. In a sense, we live right on top of a vast nuclear reactor. Can't we use all that energy? We do, to some extent. It's called geothermal energy, and it's conveniently used for space heating in some places. Using it to generate power is more difficult. The temperature of Earth's interior rises with increasing depth, typically reaching the boiling point of water at a depth of about three miles. There are only a handful of places on Earth where a geothermal source rises to within drilling distance of the surface and thus can be used for power generation. Even in those places, using the heat to generate power often cools the source faster than its heat can be replenished. Geothermal energy will always be useful but probably never a major contributor. We should remember though that without geothermal energy we would never have had fossil fuel.

There is a cheap, plentiful supply of energy available for the taking—and, like geothermal energy, this one won't run out for billions of years. It's called sunlight. We now make very poor use of the sunlight that arrives at Earth. Farmers use it to grow food and fibers for textiles. A little bit is collected indirectly, in the form of hydroelectric and wind power. Here and there, solar cells provide energy for one use or another. But by and large, the solar energy that isn't reflected back into space just gets absorbed by the earth. We could learn to make better use of it along the way.

Sunlight is not very intense, as energy sources go. The flux of energy from the Sun amounts to 343 watts per square meter at the top of the atmosphere, averaged over Earth's entire surface and over an entire year. By comparison, the continuous per-capita consumption of electric power by Americans is 1,000 watts. Nevertheless, the solar power falling on the United States alone amounts to about ten thousand times as much electric power as even Americans consume. Both sunlight and nuclear energy can be used to make hydrogen or charge batteries, in a number of ways. There are chemicals and organisms that evolve hydrogen when sunlight is added. Sunlight makes electricity directly, in solar cells. Electricity can also be generated by using sunlight or nuclear energy as a source of heat to run a heat engine—such as a turbine (see chapter 4)—that can generate electricity. By means of electrolysis, electricity can make hydrogen from water. There is not much reason to doubt that hydrogen or advanced batteries can serve our transportation needs. At present, nuclear technology is far advanced compared to solar for all these purposes, but that could change.

So, technically, scientifically, the means may exist to build a civilization that has everything we think we need, without fossil fuels. There may be a future for us. The remaining question is, Can we get there? And if it is possible to live without burning fossil fuels, why wait until the fuels are all burned up? Why not get to work on it right now, before we do possible irreparable damage to the climate of our planet?

ENERGY MYTHS
AND A BRIEF
HISTORY OF ENERGY

H ere are a few common myths about energy and related matters:

    ➚ The greenhouse effect and global warming are bad.

    ➚ There's enough fossil fuel in the ground to last for hundreds of years.

    ➚ Two dollars a gallon is too much to pay for gasoline.

    ➚ Oil is produced by oil companies.

    ➚ When we do run out of oil, the marketplace will ensure that it's replaced by something else.

    ➚ Nuclear energy is bad.

    ➚ We can help by conserving energy; otherwise, there'll be an energy crisis.

None of these statements is correct. Some are outright false, and others express poorly something important that is correct. To get it right, we have to understand how it all works. This is how it all works.

Nuclear reactions inside the Sun heat its surface white hot. From that hot surface, energy in the form of light, both visible and (to our eyes) invisible, radiates uniformly away in all directions. Ninety-three million miles away, the tiny globe called Earth intercepts a minute fraction of that solar radiation.

About 30 percent of the radiation that falls on Earth is reflected directly back out into space. That's what one sees in a picture of Earth taken, say, from the Moon. The rest of the radiant energy is absorbed by Earth.

A body that has radiant energy falling on it warms up until it is sending energy away at the same rate it receives it. Only then is it in a kind of equilibrium, neither warming nor cooling. In any given epoch, Earth, like the Moon or any other heavenly body, is in steady-state balance with the Sun, neither gaining nor losing energy. That is the primary fact governing the temperature at the surface of our planet.

The rate at which Earth radiates energy into space depends on its temperature. Because it receives only a tiny fraction of the Sun's energy, it radiates much less energy than the Sun does, so it can balance its energy books at a temperature much cooler than that of the Sun. In fact, it can radiate as much energy as it receives with an average surface temperature of 0 degrees Fahrenheit. Earth's radiation is not visible to our eyes and is called infrared ("below red") radiation, because its color is below the red end of what we are capable of seeing.

Fortunately for us, that's not all there is to it. If the average surface temperature of Earth were 0° F, we probably wouldn't be here. Earth has a gaseous atmosphere, largely transparent to sunlight but nearly opaque to the planet's infrared radiation. The blanket of atmosphere traps and re-radiates part of the heat that Earth is trying to radiate away. The books remain balanced, with the atmosphere radiating into space the same amount of energy Earth receives but also radiating energy back to Earth's surface, warming it to a comfortable average temperature of 57° F. That is what's known as the greenhouse effect. Without the greenhouse effect and the global warming that results, we probably would not be alive.

There is a tiny but vital exception to the perfect energy balance of the Earth-Sun system. Of the light that falls on Earth, an almost imperceptible fraction gets used up nourishing life. Through photosynthesis, plants make use of the Sun's rays to grow. Animals eat some of the plants. Eventually, animals and plants die. Natural geological processes bury some of that organic matter deep in the earth. As we have seen, that is how the fossil fuels are produced. So a tiny fraction of the distilled essence of sunlight is stored in the form of fossil fuels. Though the process of accumulating these fuels is agonizingly slow and inefficient, it has been going on for hundreds of millions of years and Earth has built up a substantial supply.

As we saw earlier, many experts think there is enough oil in the ground to last for decades and enough coal for hundreds of years, at the present rate of consumption. Among other fallacies, that view rests on the unstated assumption that the oil crisis will occur when the last drop of oil is pumped and likewise for coal and the other fossil fuels. The more sophisticated Hubbert analysis tells us that we get into trouble when we reach the halfway point. That's when the rate at which we can

extract oil or other fuels starts to decline. But that isn't the only fallacy in that rosy picture. The present rate of consumption is the biggest myth of all. For one thing, we Americans consume fuel at five times the average per capita rate of the rest of the world, and the rest of the world wants in. For another, there is a powerful inverse correlation between per capita energy consumption and female fertility. The richer the nation, the higher the rate of fuel consumption and the fewer the number of children born. If the whole world is brought up to first-world status as quickly as possible, then sometime later in the century there might be ten billion people on Earth living in relative comfort and burning lots of fuel. If, instead, the third world remains in poverty, there might be a hundred billion people on Earth living in misery, *and consuming the same total amount of energy*. Either way, all the fossil fuel will run out a lot faster than predicted by the present rate of consumption.

Americans grumble about paying two dollars for a gallon of gasoline, but visitors from Europe are usually astonished to discover that gasoline is just about the cheapest liquid you can buy in the United States. Two dollars a gallon amounts to fifty cents a liter. Americans are willing to pay twice that or more for bottled drinking water. One consequence is that we Americans, with 5 percent of the world's population, consume 25 percent of the world's oil. Cheap gasoline is not the solution; it's a big part of the problem.

All the oil pumped worldwide amounts at present to about twenty-five billion barrels annually. The oil companies refer to that as "production," but no oil company really produces a drop of oil—that's one reason it's so cheap. Of course, poking a hole in the ground and figuring out where to poke it does cost something. But can it really be true that this precious fluid that has taken the earth hundreds of millions of years to accumulate is worth nothing more than the cost of pumping it out of the ground? Or have conventional economics, property rights, and the rest somehow broken down here?

Speaking of conventional economics, economists firmly believe that when the oil starts to run out, the rising price will bring other, more expensive fuels to the marketplace. As we have already seen, the truth is a little more complicated than that. History shows that we don't react in an orderly, predictable way even to a temporary shortage of our precious gasoline. And whether we panic or not, the rate-of-conversion problem is likely to defeat us. Also, no other fossil fuel can replace the cheap oil that is the cornerstone of our civilization. And finally, if we do manage to burn up the other fossil fuels too, the consequences for our climate cannot be predicted. All in all, we clearly have a serious energy problem.

Many people fear nuclear energy, but in reality all energy is nuclear. The only energy sources we have are the Sun, which is a nuclear fusion reactor; natural radioactive elements in the earth, which keep the interior of the planet hot; and man-made nuclear reactions in nuclear fission reactors and bombs. Every other form of energy is derived from those sources.*

Of course what people really fear is not the Sun or the natural radioactivity buried in the earth but the nuclear reactors made by human beings. There have been some terrible accidents in nuclear reactors, but they are far less terrible than some of the accidents that have taken place in the history of coal mining or oil drilling. More than one hundred thousand men and boys died in the coal mines of England alone, in the second half of the nineteenth century.[1] By contrast, at Chernobyl, the only nuclear reactor accident that caused a substantial number of deaths, there were only 31 immediate deaths, and those were mostly heroic firemen who knew they were risking their lives to bring the situation under control. The total number of deaths,

---

*There is one very small exception: energy from the tides. The only existing plants are at the mouth of the La Rance river estuary on the northern coast of France, and at the Bay of Fundy, Nova Scotia. The tides are raised by the gravitational pull of the moon and the Sun, but their energy comes from the spin of the Earth on its axis.

estimated to have been about 2,500, should be compared to long-term deaths due to fossil fuels, e.g. caused by emphysema and the like.[2] While fuel for nuclear fission is also a finite resource, well-run nuclear reactors are easily the safest and cleanest source of energy that is practical at this moment, provided we can find reliable ways to dispose of the radioactive wastes that result. There is one other possibility: energy from controlled nuclear fusion. In chapter 5, we'll explore the technical differences between proposed nuclear fusion and the nuclear fission reactors of today. Nuclear fusion would use a fuel supply that is nearly inexhaustible, and we know of no scientific principle that forbids it from working. As noted, however, it has proved remarkably elusive; in spite of billions of dollars invested, nuclear fusion has been twenty-five years away for the past fifty years. It seems unwise to bet the future of our civilization either on conventional nuclear fusion or on the cold kind that a few scientists continue to pursue in spite of the skepticism of most.

Surely one way to help guarantee our future is by conserving energy. Surprisingly, however, it is not energy that we have to conserve. One of the most fundamental laws of physics says that energy is always conserved. Energy can change from one form to another, or it can flow from one body to another, but it can never be created or destroyed. We don't have to conserve energy, because nature does it for us. For the same reason, there can never be an energy crisis. That doesn't mean we don't have a problem; it just means we haven't been describing the problem in the correct terms. There is something that we are using up and that we need to learn to conserve. It's called fuel.

A BRIEF HISTORY OF ENERGY

In the eighteenth century, heat was thought to be a fluid called caloric. Just as water runs downhill, caloric could flow down in temperature from a hotter body into a cooler one. And like water, caloric was neither created nor destroyed while it flowed. To use the jargon of modern physics, caloric was thought to be a conserved quantity. The caloric theory was rigorous and quantitative. A chunk of copper at a certain high temperature contained a known amount of caloric. If you put it into a container of a known amount of cool water, you could calculate how much caloric would flow out of the copper into the water, and thereby predict with precision the temperature at which the two substances would come to equilibrium. Nevertheless, a former American colonist named Count von Rumford found the caloric theory wanting.

Benjamin Thompson was born in Woburn, Massachusetts, in 1753. Having spied and later commanded a regiment for England in the Revolutionary War, he prudently went into exile, in England and later in Bavaria. In the course of his career, Thompson promoted the use of James Watt's steam engine (more later about steam engines), introduced the potato into the common diet, invented a drip coffeemaker, and in 1791 was created Count von Rumford by the elector of Bavaria. (Rumford was the name of what is now Concord, New Hampshire, his wife's hometown.) He is best remembered today for pointing out in a scientific paper that boring out cannon barrels seemed to create quite a lot of caloric out of nothing. According to the caloric theory, that should not have been possible.

Count von Rumford's cannon barrels and many other observations would eventually blow the caloric theory out of the water. Caloric, or heat, would not turn out by itself to be a conserved quantity. Instead, it turned out to be just one of the possible forms of what we now call energy. Rumford and others during the first half of the nineteenth century tried to measure how much friction or other mechanical action would produce a given amount of heat. In a sense, what we now call the law of conservation of energy was discovered at least nine different times. When such a thing happens, credit for the discovery goes not to the first person who discovered it but to the one who discovered it last—the one who discovered it so well that it never had to be discovered again. This person's name was James Prescott Joule.

Joule, son of a wealthy brewer, was born in Manchester, England, in 1818. Largely educated at home, he went off to Cambridge at age sixteen to study with the famous chemist John Dalton. After completing his education, he returned to Manchester, where he built a laboratory in his father's house. Throughout his life, he supported his research out of his own pocket. In his most famous experiment, he arranged for a horizontal brass paddle wheel in a water tank to be turned by means of weights and pulleys. Weights of 4 pounds each descended a distance of 36 feet at a rate of about a foot per second. Then they were hoisted up again, one after another, to keep the paddle wheel spinning. This procedure was repeated sixteen times, after which the rise in temperature of the water was measured by means of a sensitive thermometer. Joule repeated the whole experiment nine different times and performed control experiments to determine the heating or cooling of the water by the atmosphere, without the churning paddle wheel.[3] From the results of those experiments, he concluded that the amount of

Work converted to heat

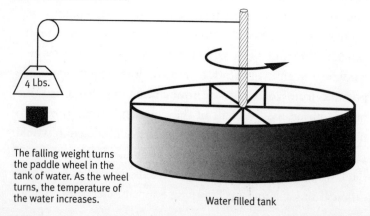

4 Lbs.

The falling weight turns the paddle wheel in the tank of water. As the wheel turns, the temperature of the water increases.

Water filled tank

heat needed to warm a pound of water by 1F°—an amount now known as the British thermal unit, or Btu—was equivalent to the amount of mechanical work required to lift a weight of 890 pounds through a distance of 1 foot. He achieved similar results in three more experiments: a magneto-electric experiment, another that involved the cooling of air by expansion, and another that measured the heating of water by constricting its flow in narrow tubes. Averaging the results of all these measurements, he arrived at a value of 817 pounds lifted through 1 foot as the equivalent of 1 Btu. The accepted value today is 775 pounds.*

It was perfectly clear to Joule that once the water had been warmed, it made no difference whether that had been done by flowing caloric from a warmer body, churning a paddle wheel, or any other means of causing mechanical friction. The warmer water had an increased quantity of something he called vis-viva. We call it energy.

The law of conservation of energy is one of the most important of all the laws of nature. Heat and work are the means by which energy can be transferred from one body or system or part of the universe to another. Energy can exist in a number of forms. Among the most important are kinetic energy and potential energy. Gravitational, chemical, and nuclear energy are all forms of potential energy.

Kinetic energy is the energy of motion. A car rolling down the street or a bowling ball rolling down the alley has kinetic energy just because it's moving. When you step on the brakes to stop your car, the car's kinetic energy is turned to heat by the friction in your brake pads. Being an ardent thermodynamicist, I own a gas/electric hybrid car. When I step on the brakes, at

*One *calorie* (a name left over from the old caloric theory) is the amount of heat needed to raise the temperature of 1 gram of water by 1 Celsius degree (1.8F°). A food calorie is actually a kilocalorie, the energy equivalent of 1,000 heat calories. Mechanical work is measured in units called *joules* (by no coincidence at all). A joule is the amount of work done lifting a weight of 1 newton (about 1/4 of a pound) through a distance of 1 meter. In terms of these units, 1 calorie has the same energy as 4.2 joules. The unit Joule himself used, the Btu, is equal to just about 1,000 joules.

least part of the kinetic energy of the car's motion goes into recharging the batteries.

Even if a body is at rest, the atoms and molecules that make it up are undergoing constant random motions and therefore have kinetic energy. The absolute temperature of a body is proportional to the average kinetic energy of its atoms and molecules, regardless of whether the body is solid, liquid, or gaseous.* The hotter it is, the faster its atoms and molecules are jiggling or bouncing around. The energy of the random motions of atoms and molecules should properly be called thermal energy, but we often loosely refer to it as heat.

Just as kinetic energy is the energy of motion, potential energy is the energy of position. For example, when Joule hoisted his weights 36 feet above the ground, he was doing work on them and thereby endowing them with gravitational potential energy. They had that potential energy by virtue of their position 36 feet above the ground. As the weights descended, their potential energy was converted to the kinetic energy of the paddle wheel and the water, and ultimately to heat, or thermal energy, in the water.

Let's follow the energy through a sequence of ordinary events. You do some work when you lift a weight off the ground. The energy to do that work came from the sugary cereal you had for breakfast. The number of food calories (or kilocalories) in each portion was written right on the box. The weight now has potential energy. You drop the weight. Its potential energy, under the influence of gravity, immediately starts turning into the kinetic energy of the falling weight. By the time it reaches the ground, the potential energy is gone, having been replaced by an equal amount of kinetic energy in the rapidly falling weight. It hits the ground

*Absolute temperature is measured in kelvins, which are the same size as degrees Celsius, but instead of placing zero at the freezing point of water, the Kelvin scale starts at absolute zero. Absolute zero is the lowest temperature possible—the temperature of a body from which every last bit of movable energy has been drained. That occurs at –273°C, or about –459°F.

with a bang, and an instant later, everything is at rest.

What happened to your work? You're not the sort of person who feels your work is unappreciated, but after all, energy is supposed to be conserved. Where did it go? The bang when it hit the floor is a good clue. That was a shock wave, propagating through the air and through the floor, both of which bounced around and eventually settled down as a slight increase in the random motions of atoms and molecules of air, floor, and everything else in the room. So the net result of the whole sequence of events is that the food energy in your breakfast cereal has turned into useless heat.

There are countless examples in nature, and in common experience, of mechanical processes that seem almost but not quite to conserve a combination of kinetic and potential energy. For example, a pendulum, as it swings through its arc, trades the kinetic energy of its motion at the bottom of its swing for the potential energy of increased height as it comes momentarily to rest at the end of its swing. But if we just set a pendulum in motion and let it swing back and forth by itself, the motion gradually seeps out of it and it comes completely to rest. We say that friction in the pivot and in the pendulum's motion through the air gradually turn its kinetic and potential energy into heat. In other words, exactly the same amount of energy we initially gave the pendulum winds up as the random, jiggling motions of atoms and molecules.

Of course, we can say that, but how can we be sure it's true? How do we know it's not just a physicist's fiction for covering up an embarrassing disappearance of energy? The answer lies in Joule's ingenious experiments. When the organized energy of the swinging pendulum turns into heat, the energy doesn't vanish without a trace. It can still be observed, in the form of a slight increase in temperature. In the case of the pendulum, or of the weight that you dropped on the floor earlier, that increase might be too small to measure accurately. But Joule devised an experiment in which the change in temperature could be measured with reasonable accuracy. Using it, he showed that a given amount of mechanical energy—that is, organized kinetic and potential energy—always turns into the same amount of heat. Thus, there is something that is conserved, always and everywhere. That something is what we call energy.

When I teach this subject to freshmen at the California Institute of Technology, there's an impressive demonstration I use. A bowling ball is suspended by a long cable from the high ceiling of the lecture hall. I stand at one end of the stage, with the bowling ball held snug to my nose and the cable stretched taut. Then I release the bowling ball to do its long pendulum swing across the width of the stage and back. As the ball goes through its long swing, there's plenty of time to remark to the class that this is an affirmation of my faith in the law of conservation of energy. If just this one time the law is violated and, after swinging 20 feet across the room and 20 feet back, the bowling ball arrives just an inch higher than it started, it will spoil my whole day. I remark, too, on just how unpleasant it is to see a bowling ball rush toward you with nothing to protect you except gravity and the law of conservation of energy. Of course it always finishes its swing not at my nose, where it started, but a few inches short, because of friction and air resistance. (It's a good idea to be careful not to lean forward when doing this demonstration.)

Energy can be stored in chemicals. Food energy is one form of chemical energy. Fuel is another. Atoms are bound into molecules, where, because of their positions relative to one another, they have certain potential energy. If they can get together with other molecules having different kinds of atoms, they might be able to rearrange themselves into different kinds of molecules that have less potential energy. The excess energy is liberated to be used for other purposes.

Because oil and natural gas are crucial to the subject of this book, let's take them as examples. Oil and natural gas are made up of hydrocarbons—that is, molecules that have varying numbers of hydrogen and carbon atoms bound together. But they are bound together rather loosely. If they are mixed with oxygen—or with air, which is 20 percent oxygen—combinations that are more tightly bound are possible. The hydrogen atoms can detach from their molecules and hook up with the oxygen to form water, which is a very stable molecule. Likewise, the carbons can combine with oxygen to form carbon dioxide, which is also very stable. When a molecule is very stable—in other words, very tightly bound—it has little or no further potential energy to give up. Because water and carbon dioxide have much lower potential energy than the fuel molecules, much energy is given off in the form of heat. That's what happens when fuel burns. The fuel won't burn, however, if it's merely mixed with air. The fuel molecules must unbind before the atoms can rebind in more favorable combinations. Some heat is needed to get the process started. That's what the spark plugs in your car are for, and that's why it takes a match to start a fire. Once burning begins, the process can produce plenty of heat to keep itself going.

The nuclei of atoms also possess potential energy. On the periodic table of the elements, all those elements lighter than iron have nuclei that can recombine into other nuclei, which are less than the sum of their parts—that is, the final nucleus has a mass when at rest that is larger than the mass at rest of any of its constituent nuclei but less than the sum of the rest masses of its constituents. The missing mass typically becomes heat, in accord with Einstein's famous formula. On the other hand, the nuclei of atoms heavier than iron can break apart, also leaving products with lower total rest mass. According to Einstein's formula, the excess rest mass of a nucleus is a form of potential energy. Iron nuclei have the lowest nuclear potential energy of any element on the periodic table.

When light nuclei combine, the process is called *fusion*. Fusion reactions provide the energy of the Sun and also the hydrogen bomb. That's the kind of nuclear reaction we have not yet learned how to make use of in reactors on Earth. When heavy nuclei disintegrate, that process is called *fission*. That was the type of reaction that took place in the earliest nuclear bombs, and it is the kind used in all man-made reactors that are currently practical.

The use of electricity is so important in our lives that no discussion of energy can be complete without discussing electrical energy. Finally, there is one other form of energy that is crucial to our story: radiant energy, the form in which energy reaches us from the Sun across 93 million miles of empty space. We turn to those subjects and to Earth's climate in the next chapter. ⊕

ELECTRICTY AND
RADIANT ENERGY

I n Europe in the eighteenth century, a hardy group of self-styled "electricians" began studying the mysteries of static electricity. They would produce various electrical phenomena, mostly through friction, and judge the strength of the electricity they produced by feeling the (usually feeble) tingle that resulted. That measurement technique suddenly became dangerous when Pieter Van Musschenbroek, at the University of Leyden in 1745, accidentally invented the Leyden jar. This device, the world's first capacitor, could store electricity, which got stronger and stronger as the friction engines of the day pumped it full. Musschenbroek's electrician colleagues, obliged to try out the new device using their time-honored measurement technique, were in for quite a shock.

They were in for an even bigger shock when the first viable theory of electricity arrived from an outpost called Philadelphia, in the vast American wilderness across the Atlantic Ocean. Although self-educated, Benjamin Franklin was one of the leading scientists of the day. In one famous experiment, he caught lightning with a kite—a stunt that had already killed at least one other person. Franklin, who knew what precautions to take, got the lightning to charge up a Leyden jar, thus proving that lightning was ordinary electricity. The English

chemist Joseph Priestley called Franklin's discovery "the greatest, perhaps, that has been made in the whole compass of
philosophy, since the time of Sir Isaac Newton."[1] Thanks to
his practical good sense, Franklin made immediate use of his
discovery by inventing the lightning rod, which was an event
of enormous importance. Before Franklin, towns in the path
of a thunderstorm would ring the church bells in the futile
hope of warding it off. Since the church steeple was usually
the best lightning rod in town, lightning would typically race
down the damp bell cord and dispatch the bell ringers. With
Franklin's invention, bell ringers were much better off and
humankind had finally found a way to control one of the
mighty forces of nature. But those achievements were by no
means Franklin's only contributions to the science of electricity. He also worked out the theory that arrived in Europe so
surprisingly from Philadelphia.

In Franklin's theory, electricity was a fluid and each material
body held a proper amount of it. If the body had more than the
proper amount, Franklin would say it was "positively charged"
with electricity. If it had too little, he would say it
was "negatively charged." To support his theory, he
devised an elegant experiment in which a small cork
with a conducting surface discharged a Leyden
jar. Suspended by a silk thread, the cork
would absorb a little electricity from the jar's
positively charged electrode, which would
then repel it in the direction of the negatively
charged electrode, where it would deposit its
excess electricity. Now negatively charged,
the cork would be attracted back to the pos-
itive. This process would continue until both had their "proper
amount" and the Leyden jar was thus discharged.

Franklin's one-fluid theory of electricity has not survived, but
his terminology has. We know today that there are two kinds of

electricity, called *positive* and *negative charge*. There is a force of attraction between positive and negative charges and a force of repulsion between like charges (either positive-positive or negative-negative). The strength of the force diminishes with distance from the charge, just as the intensity of light from the Sun diminishes with distance from the Sun. At the most fundamental level, that is all that can be said about electricity. We don't know why any of those statements are true; we just know that they are.

All matter is made up of atoms and each atom consists of a nucleus, which carries positive electric charge, orbited by electrons with an equal and opposite amount of negative electric charge. When atoms combine to form molecules, they typically share their outermost electrons, which orbit the entire molecule. Metals are made of atoms from the part of the periodic table of the elements which have loosely held outermost electrons. When these atoms combine in huge numbers to form hunks of metal, their outermost electrons become free to move throughout the material just as if it were one giant molecule. A typical metal is thus a solid filled with a fluid of movable negative charge. Not so different, really, from Franklin's theory.

Starting at the dawn of the nineteenth century, an astonishing series of inventions and discoveries were made concerning electricity. In the very year 1800, in Italy, Alessandro Volta invented the electric battery. Then, in Denmark in 1820, Hans Christian Oersted discovered that a metal wire connected to the poles of Volta's invention could deflect a compass needle. In the history of physics, great events tend to be seen as unifications; that is, phenomena that previously seemed completely unrelated turn out to be different aspects of the same thing. Isaac Newton's mechanics may have been the first great unification: The laws that govern the motions of bodies on Earth turned out to be the same as those that accounted for the orbits of the planets in the heavens. Oersted's discovery is of this class. Two phenomena, electricity and magnetism, that had previously seemed unrelated turned out to be aspects of the same thing.

When a metallic conductor is connected to the poles of a voltaic cell (another name for Volta's battery), the fluid of electrons inside the metal is pushed through it, just as the force of gravity can make water flow through a pipe. Electrons leave the metal at one pole of the battery at the same rate they enter the metal at the other end, so the conductor remains electrically neutral, always with the same amount of positive and negative electric charge. The electric force is thus neutralized and plays no role, but the flowing electrons inside the metal do produce a force that can deflect a magnetic compass needle. Electric charge at rest produces the electric force. Electric charge in motion produces the magnetic force. That was the meaning of Oersted's great discovery.

In the decade after Oersted's discovery that an electric current produces magnetism, many scientists tried to discover the reverse effect: that is, how to use a magnet to create an electric current. That discovery was made in 1831 by a remarkable English chemist named Michael Faraday.

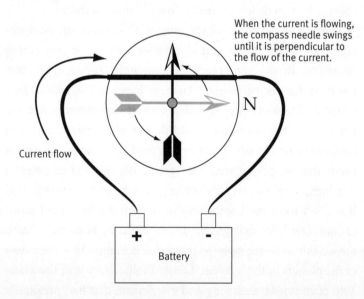

When the current is flowing, the compass needle swings until it is perpendicular to the flow of the current.

N

Current flow

+ −

Battery

Faraday, like Ben Franklin, was largely self-educated. Working in London as a bookbinder's apprentice, he was enthralled by public lectures given at the Royal Institution by the charismatic chemist Humphry Davy. (The Royal Institution, which survives and presents public lectures to this day, was founded by our old friend Count Rumford of the hot cannon barrels.) Faraday quit his job to become Davy's assistant, and eventually rose to become professor at the Royal Institution and perhaps the greatest scientist in Europe.

Faraday discovered that if a magnet was moved through a copper coil, it would induce a surge of electric current. This phenomenon is called electromagnetic induction, and it is the basic means by which most electricity is generated today. A turbine (see chapter 4) is used to rotate a coil of wire in a magnetic field. An oscillating current is induced that can be sent far away to light our lamps, run our refrigerators, and power our computers. According to legend, when the prime minister visited Faraday's laboratory to witness electromagnetic induction, he asked, "But of what use is it?" According to one

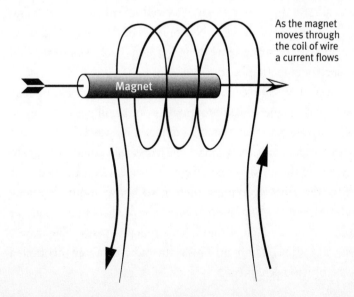

As the magnet moves through the coil of wire a current flows

Magnet

version of the legend, Faraday replied, "Of what use is a newborn baby?" The other version has him replying, "I don't know, sir, but I am sure you will find a way to put a tax on it."

In the second half of the nineteenth century, a Scottish theorist named James Clerk Maxwell produced the next great unification. He wanted to probe deeper into the connection between electricity and magnetism. A certain amount of electric charge would give rise to an electric force of a certain measurable strength. That strength was simply an empirical fact; it could be measured but not explained. If the same amount of charge was set in motion, it produced a magnetic force; that force, too, could be measured but not explained. Maxwell devised a way to compare the strength of the electric force with the strength of the magnetic force in such a way that the ratio did not depend on the details of any one experiment; that is, the ratio did not depend on the amount of charge or how fast it was moving or how far away it was. The result would have to be a universal, some value that was the same everywhere and always—a fundamental constant of nature. And what popped out when Maxwell found that magic ratio was . . . the speed of light! Not only were electricity and magnetism aspects of the same thing but light itself was some kind of electromagnetic phenomenon.

Maxwell assembled everything that was known about electricity and magnetism and found that one small piece was missing. When he supplied the missing piece, he was left with what we today call Maxwell's equations. Maxwell's equations constitute one of the most perfect scientific theories ever devised. The theory describes everything there is to know about electricity and magnetism. Among other things, it predicts that whenever electric charges jiggle around, they radiate energy that travels away at the speed of light. This is the radiant energy mentioned in the previous chapter.

If they jiggle at just the right frequency, what they radiate is visible light. But energy is also radiated at higher or lower frequencies, which we can't see; for example, gentle oscillations produce long waves that we call radio waves. As soon as Maxwell's prediction was known on the Continent, Heinrich Hertz in Germany built an apparatus that could generate and detect radio waves. The result was the triumphant confirmation of Maxwell's theory and led directly to the invention of wireless communications. At higher frequencies, but frequencies still too low for our eyes to see, the radiant energy is called infrared radiation. Moving up in frequency, there is a narrow band that we can see, called visible light, whose spectrum runs from red through yellow to violet; then, beyond what we can see, lie ultraviolet rays, X rays, and finally gamma rays. Although we give different names to these different parts of the electromagnetic spectrum, they are all the same stuff and they all propagate through empty space at the same speed, the speed of light, just as Maxwell said they should.

In the late nineteenth and early twentieth centuries, Faraday's newborn baby grew up to become a giant of technology, harnessed and exploited by (among others) another self-educated genius, Thomas Edison, and Nikola Tesla, a brilliant emigrant to the United States from Serbia. One result is that we spend our lives so immersed in electrical gadgetry that we barely notice it. However (to bring us back to the point of this book), energy, which must come from somewhere, is consumed by all electric devices. We use electricity to transfer energy from one place to another, and to do almost anything you can think of once it gets there. It most often comes from a power plant that burns fossil fuel to power a turbine generator—although sometimes the turbine is powered by heat from a nuclear reactor or by water under the pressure of a reservoir (hydroelectric power). The turbine performs work to turn the coil that induces electric current according to Faraday's principle of electromagnetic induction. The current is sent a long distance over transmission lines, to where it can be turned into more work by an electric motor, or into light by an electric lamp, or used in a myriad of other ways.

When electricity flows through a wire, it encounters resistance. All our experience tells us that any moving thing in the real world encounters resistance, in the form of friction, viscosity, and so on. Electrical resistance is of the same nature. Just as friction turns mechanical kinetic energy into heat, electrical resistance turns some of the flowing electrical energy into heat. The effect is called *joule heating,* after James Joule, who also studied this phenomenon. Because of this and related phenomena, only about 75 percent of the electrical energy generated in American stationary power plants actually gets to the end user.

It is possible to store electrical energy—for example, in the modern version of Musschenbroek's Leyden jar—but only in small, relatively unimportant amounts. One of the problems of the electric power grid is that electrical energy is very hard to store in large quantities. It isn't possible to generate power at night, when much less of it is being used, and save it for use during the day, when it's in high demand. There has to be enough generating capacity to meet the highest demand—for example, on the hottest summer day, when all the air conditioners are running—even though some of those generators will not be needed most of the rest of the time. For all practical purposes, electricity (like oil, only more so) is always consumed at the same rate it is generated. To use electricity you generally need to be connected to the electric power grid that distributes it, which means that electricity is easier to use in things that stand still, like your house, than in things that move around, like your car— a subject we'll turn to in the next chapter.

But now we're ready to take up the larger theme of radiant energy, and how it creates Earth's climate.

RADIANT ENERGY AND EARTH'S CLIMATE

In May of 1884, at the University of Uppsala in Sweden, Svante August Arrhenius, a student whose interests lay somewhere in the no-man's-land between chemistry and physics, defended his doctoral thesis. Molecules dissolved in water existed as separate, electrically charged parts, even when the solution was not conducting an electric current, Arrhenius proposed. His professors were not impressed. He was awarded a doctorate with the lowest possible passing grade. But the theory that grew out of that thesis would in 1903 make him the first Swede to receive the Nobel Prize. It was Arrhenius who, in a paper published in 1896, first put forth the hypothesis that carbon dioxide resulting from the burning of fossil fuel could lead to global warming.[2] To understand how that might happen, we have to consider the relationship between radiant energy and temperature.

We already know that all matter is made up of electrically charged electrons and nuclei that are jiggling around with random thermal energy at all times. Therefore, according to Maxwell's discovery that jiggling electric charges radiate energy away at the speed of light, all matter must be radiating away energy at all frequencies at all times. Radiation from other bodies is also falling on all bodies all the time. If the rate at which a body is radiating is not equal to the rate at which it is absorbing radiant energy, it either warms up or cools down until the two rates are equal.

Figuring out just how much a body of a given temperature radiates in each range of frequencies was a problem of truly historic proportions. German physicist Max Planck worked it out in the year 1900. He was trying to find a formula that fit measurements made by his compatriot Wilhelm Wien. He found

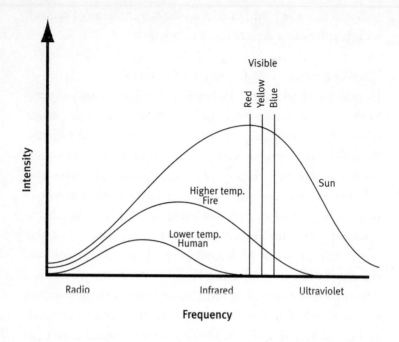

that he couldn't get agreement with the data unless he assumed that light falling on a body was absorbed not as continuous waves but as bundles of energy that we now call photons.* Planck's discovery was one of the key steps in the overthrow of Newtonian mechanics and the invention of quantum mechanics. That is not part of our story, but the formula Planck found is very much part of our story.

The sketch above shows the amount of radiation at each frequency that might be expected, according to Planck's theory, from bodies at various temperatures. At the lowest temperature, the intensity of the radiation reaches a maximum somewhere deep in the infrared and then falls back down to zero, never reaching the visible. That might be the temperature of your own body. You are radiating electromagnetic radiation at all times, but it's invisible. You don't

*In physics, individual units are generally given names ending in –ons. Thus, there are electrons and photons, the atomic nucleus is made up of protons and neutrons, and so on. There is even a theory that human populations have individual units, called persons.

glow in the dark, because you're just too cool. On the other hand, at the next higher temperature the overall intensity is higher, the maximum occurs at a higher frequency, and there is some radiation in the visible range. The visible radiation is highest at the red end of the visible part of the spectrum, so to our eyes a body at that temperature would appear to glow red. That might be the temperature of coals glowing in the fireplace. At the temperature of the surface of the Sun, the maximum radiation occurs near the visible. We see all colors simultaneously, and the result looks white. By no accident at all, our eyes have evolved to be most sensitive to the color of sunlight as it reflects from the objects around us.

Now we're ready to understand the thermal behavior of the Sun-Earth system. Nuclear reactions inside the Sun heat its surface white hot, to a temperature of nearly 6,000 kelvins, or about 11,000°F. From the surface of the Sun, radiant energy streams outward in all directions at the speed of light. We can think of this energy either in the form of electromagnetic waves or in the form of packets or bundles of energy called photons. For our purposes, the two descriptions are equivalent, and both are correct. Ninety-three million miles away, Earth in its orbit intercepts a tiny fraction of this solar radiation.

Earth, like any other body, radiates energy at all times. If it radiates less energy than it receives, it must be warming up. If it radiates more than it receives, then unless it has its own internal energy source, it must be cooling down. (Actually, Earth does have an internal source: Natural radioactivity—that is, the gradual decay of naturally occurring heavy unstable nuclei—keeps the planet's interior hot but has little effect on its surface temperature, so we will ignore that source in this analysis.) In the long term, Earth is neither warming nor cooling, so it must be radiating away just as much of the Sun's radiation as it absorbs.

The intensity of the Sun's radiation at the orbit of Earth is

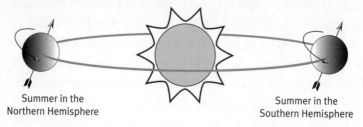

Summer in the
Northern Hemisphere

Summer in the
Southern Hemisphere

1,372 watts per square meter.* In this bath of radiant energy, Earth turns on its axis like a chicken roasting on a spit, and the energy from the Sun spreads over its spherical surface. The net result, averaged over the surface of the planet and over an entire year, is a flux of 343 watts of solar energy per square meter at the top of the atmosphere. As noted in the preceding chapter, 30 percent of that flux of energy is reflected directly back out into space, where it can be seen from the Moon as the colors of the clouds, the polar ice caps, the land masses, and the oceans.** The rest is absorbed and eventually re-radiated as invisible infrared radiation. In order to radiate, on the average, 70 percent of 343 watts per square meter back out into space, Earth's surface would have to have a temperature of 255 kelvins, which corresponds to –18°C, or 0°F. If that were the whole story, the human race, to say nothing of plants and animals as we know them, would undoubtedly not have evolved. But that is not the whole story.

Aside from the fundamental inalterable fact that Earth is 93 million miles from the Sun, a number of other factors influence the climate of our planet. Among these are its tilted axis, the El Niño cycle, the jet stream and various currents in the oceans, and the greenhouse effect.

Earth makes one complete, nearly circular orbit around the Sun each year. All of the planets in the solar system orbit the

---

*A watt is a joule per second. A Btu is about 1000 joules. So, the solar flux through each square meter at Earth's orbit is a little more than enough to raise the temperature of one pound of water one degree Fahrenheit each second.
**Roughly 17 percent is reflected back to space by clouds, 6 percent by the earth's surface, and 8 percent is backscattered by air.

Sun in a single plane, called the plane of the ecliptic. Earth also rotates about its own axis once a day; the axis is not perpendicular to the ecliptic plane but tilted at an angle of 23.5°. As Earth goes around in its orbit, the axis points in the same direction, toward the distant star Polaris.* As a result, at one time in the year the northern hemisphere is more exposed to the Sun; it is then summer in the north. Six months later, the southern hemisphere is more exposed to the Sun and it is summer in the south. Thus the tilted axis creates the seasons.

The El Niño cycle is governed by the trade winds, a steady stream of air that blows across the equatorial Pacific from South America to Asia. The wind pushes sun-warmed surface water toward Asia, where it piles up and evaporates, causing the monsoon rains characteristic of the South Asian climate. However, once every three or four years, for reasons that are not known, the trade winds slacken and the warm water spreads back across the Pacific, causing droughts in Asia and Australia, storms in the Americas, and everywhere a huge perturbation of Earth's climate.

There are a number of more or less stable flows in the atmosphere and the oceans. One of these is called the Thermohaline (thermo = temperature, haline = salt) flow in the North Atlantic. Seawater gets denser as it gets colder and as it gets saltier. In the Atlantic, warm water from the equatorial region flows northward, evaporating and getting saltier as it goes and also getting colder as it mixes with cold northern waters. Finally, in the northern reaches it gets dense enough to sink to the bottom of the ocean. This sinking provides the gravitational energy to drive the whole flow, a kind of giant, slow conveyer belt. (Once seawater sinks to the bottom it will not emerge on the surface again for a thousand years!) The Thermohaline flow

---

*Actually, the direction of Earth's axis does change, very slowly. It's always tilted at 23.5°, but it makes a complete circle about the perpendicular once every twenty-six thousand years. Thus, even in the brief span of written history, Polaris has not always been the pole star. In fact, Earth's axis doesn't point precisely at Polaris now, but it will in a few hundred years.

is thought to contribute to the relatively temperate climate of Europe; there is evidence that eighteen thousand years ago, at the height of an age when ice covered much of Europe and North America, the Atlantic flow was in the opposite direction. Apparently at that time ocean temperature was fairly uniform, but the surface water became saltier, hence denser, as it flowed south. This is just one indication that vast and long-lasting changes can take place in Earth's climate.

Of all the factors that affect our climate, by far the most important is the greenhouse effect. Think of three separate systems—Earth, the atmosphere, and space, as shown below. Each system must remain in balance, losing and gaining energy at the same rate. This balance is automatic and self-correcting. If the atmosphere or the earth gets out of balance, each will warm or cool until balance is restored. From space, 343 watts per square meter arrive; part is reflected back and the remainder is

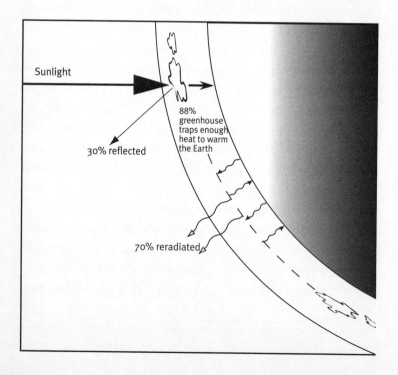

Sunlight

30% reflected

88% greenhouse traps enough heat to warm the Earth

70% reradiated

ultimately re-radiated as infrared radiation. The atmosphere, which is composed mostly of nitrogen and oxygen, is largely transparent to sunlight, so most of the radiation that is not reflected by the clouds falls directly on Earth, warming it and causing it to radiate in the infrared. The nitrogen and oxygen in the atmosphere are also transparent to infrared radiation, but the atmosphere contains traces of other gases that are *not* transparent to infrared. The most abundant of those gases is water vapor, but they also include methane, carbon dioxide, ozone, nitrous oxides, and chlorofluorocarbons. These are the so-called greenhouse gases. They absorb the infrared radiation from Earth's surface and re-radiate it, both up toward space and down toward Earth, warming our planet to an average temperature higher than it would otherwise have.

Before the Industrial Revolution, when we started burning fossil fuels in large quantities, the greenhouse gases in the atmosphere intercepted 88 percent of the infrared radiation from Earth's surface. The result of that 88-percent greenhouse effect was to raise the average surface temperature of the planet from 255 kelvins to 287 kelvins, roughly equal to 14°C, or 57°F. At this comfortable average temperature, human life evolved, climbed down from the trees, and started building steam engines. So the greenhouse effect and the global warming it produces are essential to life as we know it. However, we are now tinkering with that natural greenhouse effect. Over the last hundred and fifty years, we have increased significantly the amount of carbon dioxide in the atmosphere.

The atmospheric increase in carbon dioxide is part of a complex carbon cycle that is only partly understood. Carbon dioxide is removed from the atmosphere by dissolving in the oceans, where it may be taken out of circulation for a long time, if the water that absorbs it sinks to the bottom—for example, in the Atlantic Thermohaline flow. The dissolved carbon may also be incorporated into mineral matter that sinks to the bottom of the sea. In most places, however, surface water tends to absorb all the carbon dioxide it can and only very slowly exchanges with deeper waters capable of absorbing more.

Carbon dioxide is also absorbed by plant life, which, through photosynthesis, incorporates the carbon into organic matter. All organic molecules contain carbon. However, the carbon absorbed by plant life soon finds its way back into the atmosphere as carbon dioxide. If the plants are agricultural, some of it will be eaten. In the course of digestion and respiration, the potential energy stored in the food molecules is released by turning the carbon atoms back into carbon dioxide. If it goes into forests, the carbon dioxide is restored to the atmosphere when the wood burns or rots.

And, of course, some plant and animal matter turns into fossil fuel, from which we liberate carbon dioxide when we burn up the fuel. It turns out that a little more than half the carbon dioxide we liberate from fossil fuels lingers in the atmosphere, instead of being removed by the natural carbon cycle. Since the beginning of the Industrial Age, we have increased the amount of carbon dioxide in the atmosphere by about 30 percent.

Before the Industrial Age, the concentration of carbon dioxide in the atmosphere was about 275 parts per million. That means for every million molecules in the air, 275 were carbon dioxide. By now, the concentration is up to 370 ppm. If we were suddenly to stop burning fossil fuel, the natural carbon cycle would probably restore the previous concentration after a thousand years or so. But we are not doing that. If we go on burning fossil fuels to the point when they start to become exhausted, around the end of this century, the estimated concentration will be double the original value—about 550 ppm.[3]

The net result of tinkering in that way with the atmosphere is not easy to predict. Increasing the amount of carbon dioxide increases the amount of infrared radiation intercepted by the atmosphere and radiated back to Earth. That warms the planet slightly, causing more water to evaporate. Water vapor is a powerful greenhouse gas, so the effect of the carbon dioxide is amplified. The warming also causes the polar ice caps to shrink, reducing the amount of sunlight reflected directly back to space, which leads to even further warming. On the other hand, the extra moisture in the air tends to condense into more clouds, and clouds reflect sunlight, decreasing the warming. With the present composition of the atmosphere but no clouds, Earth would be warmer by about 7° F; thus the clouds have a huge cooling effect.[4] Also, melting the Arctic ice cap reduces the salinity and hence the density of the North Atlantic waters, which could ultimately have the effect of interrupting the Thermohaline flow and thus cooling Europe. The effect of adding CO_2 to the atmosphere is complex, with both positive and negative reinforcement acting in ways that will have consequences that are not well understood. Some scientists think that by cutting off the Thermohaline flow alone, an increase in the greenhouse effect could result in a net cooling rather than warming of the planet.

The cozy climate we now enjoy is in what scientists call a *metastable state*, in which small perturbations do not cause drastic changes. However, we know there are other possible metastable states for Earth that are dramatically different from the one we're in now. For example, suppose there were no greenhouse gases at all. The temperature would immediately drop to 255 kelvins—but that's not the end of it. At that temperature, water everywhere would freeze, reflecting more sunlight and further cooling the planet. Some geologists think that Earth went through periods like that perhaps a billion years ago. It's called the Snowball Earth theory.

On the other hand, suppose we succeed in increasing the 88-

percent greenhouse effect to 100 percent. What would the average surface temperature of Earth be then? We don't know exactly, but we do have an example to look at. The surface of Venus should of course be somewhat warmer than the surface of Earth, because of its closer proximity to the Sun; however, the difference in expected temperature is not very large and it's possible that Venus could be Earthlike. But we know it isn't. Venus has a poisonous atmosphere with a runaway greenhouse effect. When a Russian spacecraft sent a probe into the Venusian atmosphere, it recorded a surface temperature hotter than molten lead.

We don't know how big a perturbation it would take to tip Earth's atmosphere into an entirely different state, one that might make it uninhabitable. However, even the relatively small perturbations we've already made can have dramatic effects. The Arctic permafrost can soften, low-lying islands may be inundated, coastlines will change, and so on. As long as we don't understand the dynamics of the atmosphere far better than we do now, we toy with it at our peril.

Heat Engines and Entropy

Sometime in 1764, a twenty-eight-year-old Scottish instrument maker named James Watt, while trying to repair a steam engine, got the idea that you could make a better steam engine just by condensing the steam outside the cylinder where the work got done. Watt's better steam engine is the invention that kicked off our fossil fuel binge. Let's see how it works:

As the illustration at the top of the following page shows, the process starts with the burning of fuel to heat water in a boiler. As in a pressure cooker, the hotter the water the higher the pressure of its vapor. So the heat from the burning fuel is used to produce steam at high temperature and high pressure. The hot pressurized steam is allowed to pass through valve *A* into the cylinder, where it pushes out a piston. Pushing the piston out is a form of work. The piston can be linked to a flywheel, as shown in the figure, or it can be linked directly to the wheel of a locomotive. The flywheel's circular motion can be used to drive a weaving loom or the lathes of a machine shop; the locomotive can take us from London to Manchester. When the fuel is burned, it creates heat. When the piston is pushed out, work is done. Just as we saw in chapter 2 that work can be turned into heat, we see now that heat can be turned into work.

Heat converted to work

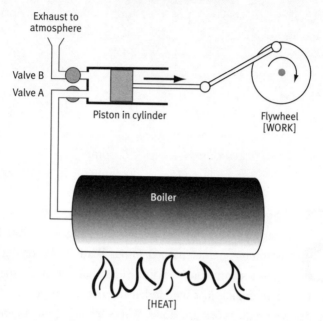

But one stroke is not enough. For the work to be really use-ful, we have to be able to push the piston out over and over again. If, once the steam has pushed the piston out, we leave things as they are, the inertia that keeps the flywheel going will push the piston back in against the pressure of the steam. Then the reverse force of the steam will quickly bring the flywheel to rest, and the engine will stop. Instead, when the piston has been pushed out, a mechanism closes valve *A* and opens valve *B*. The steam trapped in the cylinder is simply shoved out into the unresisting atmosphere. Then the two valves switch back to the original configuration, and the whole process starts all over again. Now we have a working steam engine.

The steam engine we have just designed shows that it's pos-sible to take heat from burning fuel and turn it continuously into work. We know from James Joule's experiment that a given amount of work always turns into the same amount of heat. Does the principle work the other way around? Does a given amount of heat always turn into the same amount of work? The answer is no! Just how much work can be gotten from a given amount of fuel depends to some extent on the cleverness of the engineer in optimizing the design of the engine. But no matter how clever the engineer, nature puts a limit on how much work can be gotten out. That limit is always less than, never equal to, the amount of heat energy produced by burning the fuel. Work can be turned completely into heat. But heat cannot be turned completely into work. The discovery of that fundamental asymmetry helped lead to the science of thermodynamics.

The earliest steam engines date back to the late 1600s. In the engine Watt was repairing, the force of pressurized steam was used to push out the piston, but then the whole cylinder had to be cooled down to recondense the steam and pull the piston back in. Then the whole thing had to be heated up again for the next stroke. This worked, after a fashion, but it burned enormous amounts of fuel and it was slow. In a word, those early engines were inefficient. They could be (and were) used, for example, to pump water out of mines, but they consumed far too much fuel and worked far too slowly to drive a locomotive, say, or a steamship. The entire subsequent history of the steam engine can be thought of as an endless quest for greater efficiency. James Watt, who is usually credited with inventing the steam engine, really only invented the idea that the steam could be recondensed outside the cylinder without actually removing heat from the cylinder walls. That was the invention that made the steam engine a practical device.

Just to get the idea, let's play engineer for a moment and try to improve the efficiency of the steam engine we designed earlier. The objective is to get as much work as possible out of the heat we put in by burning the fuel. The heat serves only to turn water into steam at high pressure and high temperature. So we want to make the best use we can of the temperature and pressure of the steam. In the engine cycle, the hot pressurized steam was used to drive the piston all the way out, before valve A was closed and valve B opened. When that happened, all the hot pressurized steam trapped in the cylinder was released to the atmosphere. Being hot and pressurized, it was capable of doing more work, but we've thrown that potential work away. We should be able to do better.

Suppose, instead of closing valve A when the piston is at the end of its stroke, we close valve A when the piston is only partway along. We also keep valve B closed at this point. Now a certain quantity of hot pressurized steam is trapped in the small volume of the cylinder. The steam's pressure continues to push the piston out, but since the steam is no longer being replenished from the boiler, as it pushes the piston out and the volume containing the steam increases, the steam's pressure decreases. Technically, this is called the *adiabatic stroke*; the word "adiabatic" means "isolated from sources of heat." Because we've closed valve A, the steam is now isolated from its source of heat, the boiler, so as it pushes the piston farther out, it expands adiabatically.

When the steam expands adiabatically and pushes the piston out, it is doing work. Work is a form of energy, and energy is always conserved, so something must be giving up that energy. The only source of energy available to the isolated steam in the cylinder is its own internal, thermal energy—in other words, its high temperature, which is really just the high kinetic energy of the molecules of steam in their random motions bouncing off the piston, the walls of the cylinder, and each other. Pushing the piston out, the molecules of steam slow down a bit, losing some of that kinetic energy. So as the steam expands to lower pressure, it also cools to lower temperature. Then, at the end of the stroke, when valve B opens and the flywheel pushes the spent steam out into the atmosphere, the steam will already have given up much of the high pressure and temperature extracted from the fuel and turned it into additional work. By introducing an adiabatic stroke, we have created a more efficient engine.

There is one inefficiency we will not be able to overcome, however, no matter how cleverly we design the cycle. When the spent steam rejoins the atmosphere, much of it will recondense into the water it started out as. When water evaporates, it cools (extracts heat from) its surroundings. (Just think of what it feels like to climb out of a swimming pool on a cool day.) Condensation is the reverse process. Condensing steam into water gives off heat. So part of the heat we got from the burning fuel turns into work driving the flywheel, or locomotive, but no matter how cleverly we design things, at least part of the heat must be dumped into the atmosphere to recondense the steam. It is now heat at much lower temperature, typically the temperature of the air rather than of the burning fuel, but it is still part of the energy we started out with in the fuel.

The steam engine is only one of a class of machines called *heat engines*. All heat engines start with heat at high temperature, turn part of that energy into work, and must, like the steam engine, dump part of the heat at low temperature in order to return to their starting points and keep on going. Heat engine types include the internal combustion engines of most cars (known to engineers as the Otto cycle), diesel engines, turbines, and the Stirling engine. Life as we live it would not be possible without heat engines.

In the Otto cycle—named after its inventor, Nikolaus Otto—intake air mixed with gasoline vapor is drawn into a cylinder and compressed by an adiabatic compression stroke of the piston. The adiabatic compression heats the mixture, but if the fuel is high-quality gasoline, it doesn't ignite. The fuel-air mixture is then ignited by a spark plug. The hot pressurized gas that results from the explosion pushes the piston out in an adiabatic power stroke. Then the spent gas must be cooled (that's what the radiator is for) and exhausted to the atmosphere. The octane rating of gasoline measures the extent to which it resists being ignited prematurely during the compression of the intake mixture. If your car doesn't have a high-compression engine, high-octane premium gasoline won't do it a bit of good.

The diesel cycle—also named after its inventor, Rudolf Diesel—is almost the same, except there's no spark plug. Air alone is drawn into the cylinder, and the adiabatic compression stroke heats it enough to ignite the subsequently inserted fuel. Diesel fuel, essentially identical to home heating oil, is easily ignited.

Turbine engines use a fan to compress the intake air, add and burn fuel to increase the temperature and pressure, and then drive the hot pressurized gas through a second fan designed to provide power by turning the thrust of the gas into rotary motion. There are many variations on this theme. In power plants where turbines are used to generate electric power, the fuel may be burned externally rather than internally. In the engines of jet airplanes, the power fan robs only a little of the thrust of the hot pressurized gas, the rest being used to propel the aircraft. In a fighter jet, part of the fuel may be injected after the power fan, where an afterburner provides more thrust for better performance, but much lower efficiency.

The Stirling engine is the darling of all thermodynamicists. It is an external combustion engine, in which the internal working fluid is an ordinary gas, often just air. If one side is heated and the other kept cool, the internal mechanism pushes the air around and makes a piston do work. Of all the engine types we've looked at, the Stirling is the only one that (in principle, at least) can attain the maximum efficiency permitted by nature. It was invented by the Reverend Robert Stirling in 1816, before Sadi Carnot, a young French military engineer, showed that there *is* a maximum efficiency permitted by nature. The Stirling cycle has been used to make special refrigerators that operate at very low temperatures, but Stirling engines that could produce enough power, say, to propel a car have not yet proved practical.

To a thermodynamicist, all these engine types are variants of the same thing: They are just heat engines, represented symbolically on the following page. They all take heat from a source at high temperature, turn part of that heat into work, and disperse the rest at low temperature. Incidentally, a refrigerator is a heat engine running backward, as also shown on the following page. Work is put in (by the electric motor that drives the compressor), heat is extracted from the cooling food inside, and the energy of the heat plus the work is deposited as heat at higher (usually ambient) temperature. That's why your refrigerator warms your kitchen.

Electric motors are not heat engines. They don't need to dump waste heat at ambient temperature, so they can be, and often are, close to 100 percent efficient. Of the energy available in electric form, nearly all can be turned into mechanical work. Electricity is easy and convenient to use in stationary applications, where it can flow right out of the national power grid. Batteries make it possible to use electric energy in mobile applications, such as cell phones and flashlights, where not too much energy is needed. A battery consists essentially of two different metals in a bath of chemicals that provides the energy to drive an electric current through an external circuit. The amount of chemical energy stored in a given volume of battery chemical, however, is tiny compared to the amount stored in the same volume of gasoline. That's the principal obstacle to electric cars.

Prodded by California law to manufacture zero-emission vehicles, a condition that can be met only by electric power, some automakers came out with versions of electric cars. The General Motors version, the EV1, was a sleek vehicle that maximized its range by dint of lightweight construction and aerodynamic styling. The manufacturer claimed that the 1997 model had a range of 50 to 100 miles between charges, but when the

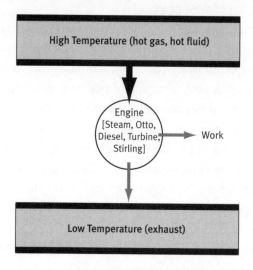

Heat engine running forward

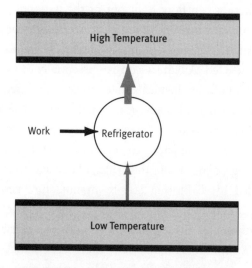

Heat engine running in reverse

batteries were freshly charged the on-board computer begged to differ, announcing that another multihour charging session would be needed after thirty miles. The EV1 was distributed only in California and Arizona, because it didn't work very well in cooler climates. It was not a practical substitute for the family car. General Motors has now pulled the plug on the EV1, and Ford has similarly killed its electric car program. The original EV1 used the same kind of lead-acid batteries that gasoline-powered cars use for their electric circuits, and the final version used somewhat better nickel-metal hydride batteries. Under pressure to make smaller, lighter cell phones, engineers have recently developed a different kind of battery, called a lithium-ion battery, that stores five times as much energy as a lead-acid battery in a given volume. That's still tiny compared to the wallop packed by gasoline, but a lightweight, well-designed vehicle with five times the range of an EV1 starts to resemble a believable means of transportation.

A different hope for the future of electric vehicles is the use of fuel cells rather than batteries. A fuel cell acts like a battery, except that it uses flowing fuel rather than discharging and needing to be recharged. In the version most likely to be used in cars, the fuel is compressed hydrogen gas. The hydrogen flows into one side of the cell and air flows along the other. Each hydrogen molecule gets separated by a catalyst into electrons and protons. The electrons are left behind, forming the negative electrode, while the protons diffuse through a special membrane. On the other side of the membrane, the protons react with oxygen from the air to form water vapor. That process consumes electrons, making this the positive electrode, and liberates energy. The electrons make the trip from negative electrode to positive electrode through an external circuit, running the car's electric motor along the way. This is called a PEM (proton exchange membrane) fuel cell. It requires expensive, precious metal catalysts, and complex construction. Although it is not quite ready for prime time yet, much research is being devoted to make it a practical device.

So the fuel cell is one possibility for the future of automotive transportation. However, the fuel it uses, hydrogen, is not a source of energy. Instead, it is a means of storing and transporting energy. The primary energy to make hydrogen will most likely be electricity produced by a heat engine using some other fuel. That kind of conversion of energy from one form to another always involves inefficiencies. Using present-day technology, to extract a given amount of hydrogen from water and compress it for use in an automobile would require consuming in heat engines about six times as much fuel as the hydrogen would replace.[*]

Most electric power is generated by heat engines in power plants; thus electric motors can generally be thought of as surrogate heat engines, doing locally the work that was generated by a heat engine somewhere else. Electric motors as surrogate heat engines work well as long as they're attached to the national grid, but they don't work as well for vehicles. Collectively, heat engines are the motive force behind our civilization. They propel our cars, trucks, trains, ships, and planes. They refrigerate our food and cool our homes and workplaces in summer.

The steam engine played a crucial role in getting us to where we are today. It kicked off the Industrial Revolution, fostered the discovery and development of fuels, and endowed us with immense powers. But it also had another, more subtle effect. Today, advances in technology are generally the result of scientific discoveries, but with the steam engine it happened the other

---

[*]According to Alec N. Brooks in his paper "Perspectives on Fuel Cells and Battery Electric Vehicles," presented at the California Air Resources Board workshop on zero emission vehicles, December 5, 2002, 1 kilogram of compressed hydrogen will drive a car about as far as a gallon of gasoline. A heat engine can produce about 10 kilowatt-hours of electric energy from a gallon of gasoline. To make and compress 1 kilogram of hydrogen requires 60 kilowatt-hours of electric energy, the equivalent of six gallons of gasoline. Just by way of comparison, a lead-acid battery stores .03 kilowatt-hours of electric energy per kilogram of battery, equivalent to about 1/300 of a gallon of gasoline.

way around. It was developed without much understanding of science, but reasoning about the steam engine by Sadi Carnot led to one of the most profound scientific discoveries of all time.

THE ENTROPY PRINCIPLE

Nicolas-Léonard-Sadi Carnot was born in 1796 in Paris. His was a prominent family of French politicians, soldiers, and engineers. The remains of two Carnots—both members of the family but neither of them our Sadi—are interred in the Pantheon. Sadi entered the elite École Polytechnique in Paris at the minimum age of sixteen, and after graduating went on to a two-year course in military engineering at the École du Génie at Metz. These were turbulent times. The fortunes of his father, a minister in Napoleon's government, rose and fell as Napoleon triumphed, was deposed, returned to power, and was deposed again after the famous reign of a hundred days in 1815. Hampered by the tribulations of his well-known father, Sadi's military career went nowhere. In 1819, at the age of twenty-three, he went into semi-retirement and devoted himself to his other interests.

Carnot was profoundly disturbed that unlettered British engineers had figured out how to make better steam engines while the French lagged far behind. Surely with his superior erudition he could do better. As it turned out, he made no contribution at all to improving the technology. He didn't concern himself with real, hissing, clanging steam engines. He didn't even assume that an engine needed to use steam. Instead he tried to abstract the essence of what an engine was all about. In the course of working that out, he invented the beginnings of a powerful new science.

The preceding discussion of the steam engine would have mystified the working engineers of the early nineteenth century who actually figured out how to make them. Theirs was a rough and ready craft, based largely on trial and error. They figured out that an adiabatic power stroke would make steam engines more efficient, but the law of conservation of energy and its corollary, the idea that heat and work were forms of the same thing, were far in the future. They were admirable and successful engineers, but they were not theorists. A theorist is just what Sadi Carnot was.

If Carnot had known about the conservation of energy, he might have assumed that the business of making a more efficient engine consisted of trying to turn 100 percent of the fuel energy into work. If so, he might not have made his important discoveries. Instead, for Carnot, heat was caloric—a fluid that could not be used up or turned into something else, such as work. So, how could caloric make an engine run?

Carnot reasoned by analogy to a waterwheel. Running downhill, water can do useful work by turning a waterwheel. The water doesn't get used up; it merely descends to a lower level. There's still just as much water at that level, but it can no longer be made to do useful work. Just so, thought Carnot, caloric can do work while running through an engine from high temperature to low temperature. At the end, the caloric is still there; it's just not as useful anymore, because it can no longer do work. To translate Carnot's reasoning into modern terms: Heat energy at high temperature is capable of driving an engine and doing work. The same amount of energy at low temperature is not capable of doing useful work. The same amount of energy is still there, but something about it has changed. That something is what we have come to call *entropy*.

Carnot then posited the reasonable principle that it is not possible by any means to have the net effect of making caloric run backward, from low temperature up to high temperature. If that could be done, it would be possible to get something for nothing. The high-temperature caloric that had been created could be used to drive an engine without using up any fuel. Carnot made the very sensible engineering assumption that it's impossible to get something for nothing. The reasonable principle that Carnot has assumed here is the second law of thermodynamics. The first law of thermodynamics says that energy is always conserved. The second law tells us why that won't do us much good when the fuel starts to run out. Carnot didn't discover or prove the second law, he simply assumed it. But that's the nature of the second law. It's an axiom whose verification lies in the fact that all predictions made on the basis of it always turn out to be true. And they do. Always.

Since the second law is so important, let's look carefully in modern terms at what it says. It says you can't build a machine that will extract heat at low temperature and deposit that same amount of energy at higher temperature, without having any other effect. "But," you say, "that's exactly what the refrigerator in my kitchen does. It extracts energy from the cooling food inside, and dumps that energy at higher temperature into my kitchen." Yes, that's right. But in order for your refrigerator to function, there must be another engine somewhere (most likely an electric power plant, far away) that is burning fuel to create heat at high temperature, part of which is turned into work in the form of electricity, which drives your refrigerator. The principle doesn't say that somewhere along the way some heat can't be forced uphill, as in your refrigerator. It says instead that *the net effect* of the power plant and your refrigerator running in tandem must always be to make heat run downhill from high temperature to low. In other words, to drive your refrigerator the power plant must always make more heat run downhill than the fridge makes run uphill. That principle is never violated.

Sadi Carnot published only one work in his entire life—a book in 1824 whose title, translated from the French, is *Reflections on the Motive Power of Fire and the Proper Machines to Develop That Power*. The book was well received, reviewed with praise for the clarity of its arguments and theorems, and then quickly forgotten. Carnot himself died in a cholera epidemic in 1832, at the age of only thirty-six. His *Réflexions* did not remain in obscurity, however. It was rescued by the engineer Émile Clapeyron, who in 1834 published his *Force Motrice de la Chaleur*, reformulating Carnot's arguments; in that form, they would become the basis of the new science of thermodynamics. In spite of using the incorrect caloric theory, Carnot's reasoning was so careful and rigorous that his results concerning the efficiency of engines are still routinely used by engineers to this day.

In Carnot's world and ours, heat always runs downhill, from high temperature to low. Just as all the rainwater that falls on the land eventually finds its way to the ocean, all heat, generated by whatever means, at high temperature will eventually wind up as the same amount of heat at ambient temperature. If we devise a "proper machine," part of the heat energy can be made to do work along the way, but whatever the case, it always winds up in the air or the oceans. Of course, "high temperature" and "low temperature" are relative terms. In principle, a machine can be made to do work using heat as it runs between any two temperatures, as long as one is higher than the other. But Carnot was able to show that the higher the high temperature, and the lower the low temperature, the more efficient the engine would be. For all practical purposes, the burning fuel determines the high temperature and our surroundings—the air or the seas—become the low temperature for real engines.

Rudolf Clausius, a German professor of mathematical physics, combined Carnot's law of heat running downhill with Joule's law of conservation of energy into the two main laws of thermodynamics as we know them today. Heat is energy, and so is work. Energy is always conserved, but when it's at high temperature—and, for that matter, when it's still in the form of unburned fuel—it's capable of being turned at least partly into work. The same amount of energy, dumped into the atmosphere at ambient temperature, has become useless. The quantity of energy hasn't changed but somehow its quality has changed. How to describe this change of quality?

To do that job, Clausius coined the word *entropy*. Entropy measures the temperature of energy. A given amount of thermal energy—the random motions of atoms and molecules—has low entropy when it's at high temperature, and the same amount of energy has higher entropy when it's at lower temperature. Thus the principle that energy always runs downhill to lower temperature is completely equivalent to saying that entropy always increases.

Students taking physics often have difficulty understanding the distinction between temperature and heat. The English language doesn't help much. When the temperature rises, we say it's getting hotter. But heat and high temperature are not the same thing, and the distinction is the crux of the issue we are dealing with here.

Heat is energy, and with energy, size matters. With temperature, it doesn't. If a body has a certain temperature, the amount of thermal energy (heat) it contains depends on how big it is. On the other hand, the temperature of a body does not depend on how big the body is. A teaspoon of water can have the same temperature as the ocean. Think of heat and the temperature of two different gases. One is the gas involved in the combustion of a fossil fuel; it could be the air in the combustion chamber of a steam engine. Heat is the quantity of energy that comes from the burning fuel. Once the fuel is burned, it takes the form of the kinetic energy of the gas molecules bouncing rapidly around in the combustion chamber. The absolute temperature is directly proportional to the kinetic energy of the gas molecules. So far there seems to be little difference between heat and temperature. Eventually, however, the same quantity of heat will wind up at lower temperature, the temperature of ambient air. Then, too, it takes the form of kinetic energy—in this case, a very slight increase of the kinetic energy of the air molecules in the atmosphere. On the average, each molecule of the cool gas (the ambient atmosphere) has gained much less kinetic energy than each molecule of the hot gas (the air in the combustion chamber) gained from the burning fuel. But the total amount of energy is the same. That must mean that the same quantity of heat is spread out into a much larger quantity of gas. Instead of being concentrated in the intense heat of the combustion chamber, it has spread into the benign cool of the great outdoors. This spreading out of heat as the temperature decreases renders the energy useless to do work. That is the essence of increasing entropy.

Entropy also has a more subtle meaning, which follows from its definition as a measure of the temperature of energy. Imagine two bodies at different temperatures. Between them, they have some total amount of energy. Now put them in contact with each other. Heat flows from the warmer body into the cooler, until the two bodies have the same temperature. That's called coming to equilibrium. But whenever heat flows from a warmer to a cooler body, entropy increases; thus, coming to equilibrium always involves increasing entropy.

Before the two bodies came to equilibrium, the world was, in some sense, a more organized place. When I come home after a long, hard day, I want to take a hot bath and drink a cold beer. Equilibrium—that is, equally tepid beer and bathwater— is definitely not my preferred state. But it is nature's preferred state. Thus, low entropy is associated with ordered, organized states of being, and high entropy is associated with the random motions of an object in equilibrium. The basic law of the universe is that, everywhere and always, entropy is increasing. In a celebrated paper of 1865, Clausius stated the first and second laws of thermodynamics:[1]

1. The energy of the universe is constant.
2. The entropy of the universe tends to a maximum.

In fact he could have been a little more forceful in his statement of the second law. *Every process that occurs* in the real world increases the entropy of the universe. Once any increase has occurred in the entropy of the universe, it is absolutely final and irreversible. The entropy of the universe can never be reduced.

It may seem a bit grandiose to express what started out to be an analysis of steam engines as laws about the universe. However, the second law has indeed turned out to be applicable to the entire universe. A straightforward application of it to the entire universe by physicist George Gamow in 1948 led to the prediction of the cosmic background radiation left over from the Big Bang that started it all—radiation discovered some two decades later and still studied by astronomers and physicists.

But applying the entropy principle to the entire universe wasn't the original idea. The original idea was to emphasize that in applying these laws you mustn't consider only part of what's happening. You have to consider the whole thing. You *can* reduce the entropy of one thing, provided that in doing so you increase the entropy of something else by even more. Then the two processes considered together will have increased the entropy of the universe. Remember the refrigerator in your kitchen. It can seem to violate Carnot's version of the second law (heat always flows downhill) by extracting heat from the food and depositing it in your kitchen. But that's only possible because there's a power plant somewhere that's making more heat run downhill than your refrigerator is sending uphill.

The perfect engine—the one envisioned by Sadi Carnot—would go through its cycle without increasing the entropy of the universe. Each step of his ideal engine's cycle worked equally well running backward as a refrigerator. Thus a Carnot engine could produce work from heat running downhill (that is, from high temperature to low), and the same amount of work used to drive a Carnot refrigerator would make the same amount of heat flow uphill (from low temperature to high). A Carnot engine running in tandem with a Carnot refrigerator would have no effect at all! Each step is perfectly reversible. The entropy of the universe is unchanged.

Even though the Carnot engine doesn't increase the entropy of the universe, it still has a limited efficiency. It obeys the iron rule of heat engines that only a fraction of the heat energy in the fuel can turn into useful work. The maximum possible efficiency of any heat engine (called the Carnot efficiency) depends on how high the temperature of the heat source (i.e., the burning fuel) is and how low the temperature at which the excess heat is deposited is.

All real engines do less well than the ideal engine envisioned by Carnot. Whether by the design of the engine cycle or by the ever-present effects of friction and other dissipative processes that are a necessary part of any real engine's operations, entropy happens. The rule of thumb commonly used by engineers is that only about a third of the energy content of a heat source, such as fossil fuel, can be turned into a useful form, such as electricity. Even more energy is lost because of transmission over long distances. So even though electric motors can be nearly 100 percent efficient, the combined result of a fossil-fuel–burning power plant generating electricity, transmitting the electricity over miles of transmission lines, and powering an electric motor at the end of the line is that less than a third of the energy stored in the fuel can turn into work. That is an inescapable consequence of the entropy imperative that rules the natural world.

There are a number of ways of stating the second law. Saying that heat always runs to lower temperature is equivalent to saying that entropy always increases. Perhaps the most profound interpretation of the second law is that it determines the flow of time itself. If you saw a movie of spent heat reorganizing itself into unburned oil, or for that matter, a scrambled egg unscrambling, you would know the movie was running backward. What is perhaps the most elegant statement of the second law of thermodynamics predated Carnot by almost eight hundred years. The Persian poet and mathematician, Omar Khayyam, as translated by Edward Fitzgerald, wrote:

> *The moving finger writes*
> *And having writ, moves on.*
> *Nor all your piety nor wit*
> *Shall lure it back to cancel half a line,*
> *Nor all your tears wash out a word of it.*

If the entropy of the universe were to reach its maximum, the universe would be in equilibrium and nothing more could happen—a fate famously known as "the heat death of the universe." In the decades after the discovery of thermodynamics, scientists and philosophers debated the meaning of what seemed to be a clear prediction by the most fundamental laws of physics of that terrible endpoint of existence. That debate has never really been resolved. Such a fate is far away, however. As long as the Sun and trillions of other stars go on shining, the entropy of the universe will continue to increase.

SUMMING UP

Now that we understand the basic principles of energy and entropy, let's replay the story of how it all works. The radiant energy that arrives at Earth from the Sun at a temperature of 6,000 kelvins is a very low-entropy form of heat. Some of it is reflected, making it possible to see Earth from space, and the rest is absorbed by the atmosphere and Earth itself. A tiny fraction of that energy is stored away at low entropy in the form of the high potential energy of fossil fuels. The rest drives the winds and ocean currents, causes water to evaporate and make the clouds, or simply falls as sunlight upon the land. Eventually all the energy that wasn't reflected back into space winds up as the thermal energy of the waters and the land, at the temperature of Earth's surface. The energy is absolutely conserved, but its entropy has increased irreversibly and forever. Earth warms to a temperature at which it radiates (in the form of invisible infrared radiation) just as much energy as it receives. Most of that is trapped by the atmosphere and radiated back to Earth, but a portion of it, equal almost exactly to the amount absorbed from the Sun, is radiated back into empty space, as cold, useless infrared radiation. Viewed from space, the whole drama has had no point but to do its inexorable job of increasing the entropy of the universe. ⬤

Technological Fixes

E arth's orbit around the Sun—or, for that matter, any solar orbit—is determined by a precise balance between the tendency of a body in space to fly off in a straight line and the gravitational force of the Sun, which continuously bends the line into an orbit. One result of that balance is that the amount of time it takes to make one complete circuit—that is, one year in the life of a planet—depends entirely on the planet's distance from the Sun. The closer to the Sun, the faster a body must move to stay in orbit and the shorter its year. If a body between the Sun and Earth tried to orbit the Sun at a speed that would take it around the Sun in one Earth year, it would be moving too slowly for its orbital speed to balance the Sun's gravity. Instead of orbiting, it would spiral down into the Sun.

However, there is an exception to this rule. There is one point, on a straight line from the center of the Earth to the center of the Sun, where a body could orbit the Sun in 365 1/4 Earth days without falling into the Sun. That point in space, known as Lagrangian point L1 (named for the French mathematician Joseph-Louis Lagrange) is about one-hundredth of the way from Earth to the Sun. At that distance, the outward pull of Earth's gravity on the body just balances its tendency to fall into the Sun. A body placed there with the right orbital speed would remain safely in orbit.

One of the technological fixes that have been proposed to alleviate further global warming is to place a giant parasol 1,200 miles across—big enough to block a few percent of the Sun's radiation—at L1.[1] From Earth, it would look like a permanent spot on the face of the Sun, and once in place it would free us to burn up all the fossil fuel on the planet without worrying about our climate. This grandiose scheme, an example of what is called planetary engineering, is clever, amusing, and probably technologically possible. But it's a foolish idea. For one thing, as we've already seen, the Earth's climate is not an inevitable consequence of the flux of solar radiation at the position of our orbit. With the same solar flux, Earth could be a ball of ice or a Venusian inferno. Changing the flux a bit cannot guarantee our safety. And, of course, a parasol at L1 is just about the last thing a future energy-starved world will need.

Shading the Earth with a giant parasol may not be a great idea, but if we are going to burn up all the fossil fuel we can get our hands on, limiting the damage done by carbon dioxide is an excellent one. How can that be done?

Carbon is the essential element of life. Every molecule of every bit of living matter contains carbon. Life is nourished by photosynthesis, the process by which plants consume sunlight, water, and carbon dioxide and produce the stuff of which they are made. Thus, growing plant life removes carbon dioxide from the atmosphere. Unfortunately, it doesn't remove it for long; carbon dioxide is returned to the atmosphere when plants rot or burn. One decidedly low-tech suggestion for getting some carbon out of circulation for a long time is to sink dead plant matter to the bottom of the sea before it has a chance to rot.[2] A more ambitious idea is to remove the carbon dioxide from the exhaust gases of fossil-fuel–burning power plants and sequester it somewhere. But where to sequester it?

One suggestion is to use the carbon dioxide to pressurize oil and natural gas wells. The pressure would help to push the oil and gas out and the CO_2 would stay behind in the well. Nothing wrong with that, and in fact it is being done successfully by the Norwegian company Statoil, but oil wells don't offer the necessary storage capacity.[3] We humans have already increased the concentration of CO_2 in the atmosphere by about 100 parts per million: If we had some way of extracting carbon dioxide from the atmosphere and storing it, we would need to store away about a ten-thousandth of the atmosphere of the entire planet, just to return the atmosphere to its pre-industrial condition. That's a lot of gas to put away somewhere. Even if we just try stabilizing the atmosphere, we will have to put away several times that amount before the planet's fossil fuel starts running out at the end of the century.

The only place that would have the necessary capacity is the ocean floor. In the atmosphere, CO_2 is a gas, but at the temperature and pressure of the deep oceans it would become a liquid denser than water and would sink to the bottom. Compared with the gaseous form, liquid CO_2 is very compact. It would take up a volume hundreds of times smaller than the equivalent gas, and the ocean floors cover most of the globe. There would be enough room.

The scheme of sequestering CO_2 and injecting it into the deep oceans might have some drawbacks, however. There is all sorts of life at the bottom of the oceans, and a layer of liquid carbon dioxide there might snuff it out. Moreover, some of the CO_2 will dissolve in the seawater and gradually diffuse upward. An increased concentration of CO_2 will alter the acidity of the oceans somewhat, which could be a very big headache for a lot of marine life, large and small. Also, any CO_2 that found its way to hot spots and mid-ocean ridges, where magma bubbles up, would be vaporized and might make it to the surface and escape. Finally, although liquid carbon dioxide would be physically stable on the cold parts of the ocean floor, it is even more stable as gas in the atmosphere. The liquid at the bottom of the sea is in a metastable state: Any event that stirs the ocean's waters—say, an asteroid impact—would suddenly release a large quantity of CO_2 to the atmosphere, with possibly catastrophic results.

Another idea is to sequester the carbon in the form of magnesium carbonate bricks. Magnesium carbonate is a kind of rock, and carbon is naturally sequestered that way as the rock forms, but very, very slowly. The idea would be to find ways to accelerate that process on a vast industrial scale. Of course, we are burning fossil fuels because it is cheap and profitable to do so. Unless the world suddenly develops a huge need for magnesium carbonate bricks, it's a bit difficult to see who would mount that giant effort so that others could continue to profit from selling fossil fuels. The United States government comes to mind as a possible sponsor, but it has not shown much appetite for environmental measures. In March of 2001, for example, the Bush administration announced that it would not ratify the Kyoto Protocol, which had been signed in 1997 by the Clinton administration, the European Union, and Japan. The Kyoto Protocol calls for industrialized nations to reduce slightly the rate at which CO_2 is poured into the atmosphere. The United States deemed this an unacceptable threat to its economic well-being. That may be just as well: The Kyoto Protocol might be too strong for the U.S. government, but it's too weak to do much good for the atmosphere.

NUCLEAR POWER

As we reach the end of the age of fossil fuels, the world will have to consider increasing the use of nuclear power. It may be the only proven technology capable of filling in for the loss of those fuels on a massive scale. There are difficulties and dangers associated with nuclear power, but there may be no alternative.

The atomic nucleus is a tiny speck at the heart of every atom which contains nearly all of the atom's mass. It consists of protons, which have positive electric charge and therefore hate to be at close quarters with one another, and neutrons, electrically neutral particles that help to keep the unhappy protons together. The position of an atom in the periodic table of the elements is completely determined by the number of protons in its nucleus. The number of neutrons in the nucleus of a given element is not as rigidly fixed, however. All elements can have atoms with different numbers of neutrons—and therefore different masses. These are called isotopes.

For example, the element uranium has 92 protons in its nucleus. Its most commonly found isotope has 146 neutrons, for a total atomic mass of 238—that is, it has 238 times the mass of a single proton. (Protons and neutrons have nearly the same mass.) We'll use the scientific designation ^{238}U for that isotope. Another isotope, ^{235}U (with three fewer neutrons), comprises approximately 0.7 percent of natural uranium. This rare isotope has the important property that if it captures a slowly moving free neutron, it breaks violently apart usually into two large nuclear fragments, plus (on the average) 2.43 free neutrons. A large part of the energy is in the form of the kinetic energy of those fragments, but the neutrons are pretty fast too. If the neutrons can be slowed down before they escape, they can be captured by other ^{235}U nuclei, producing more free neutrons, which in turn collide with still other ^{235}U nuclei, and so on. That is what is meant by the term *chain reaction*. The only way to slow the neutrons down and thus initiate the chain reaction is to make them collide with something not too much heavier than themselves. If they collide with a big, heavy nucleus, they bounce off it like a rubber ball bouncing off a concrete wall, retaining nearly all their energy. But if they bounce off something lighter, that object is set in motion and removes some of their energy. In most nuclear reactors, water is used for this purpose.[*] The water also serves to cool down the fuel, which is heated by slowing the energetic nuclear fragments, and that heat can be used to drive a turbine. Thus, a nuclear reactor is just a source of heat, like burning coal.

Today in the United States, about half our electricity is generated by burning coal, with nuclear power contributing about 20 percent. In other countries, the amount of nuclear power varies from most (France) to none (for example, Italy, which has outlawed nuclear power plants but buys electricity from French nuclear plants). Suppose the world was prepared (or forced) to ignore or solve all the difficulties associated with nuclear energy and forge ahead to replace all stationary fossil fuel power with nuclear power. That would certainly help relieve the carbon dioxide problem in the atmosphere. Nuclear power would not replace oil for use in transportation, but let us suppose that the electricity it produces can be used to make an alternative mobile fuel, such as hydrogen. Is there enough uranium around for that to be a long-term solution?

Estimating uranium reserves is not nearly as well developed a science (or art) as estimating oil reserves. Just like oil reserves at an earlier time, uranium reserves will surely increase, as a result of both further exploration and advancing technology. However, known reserves are estimated to be enough to supply all of Earth's energy needs—at the current rate of energy consumption—for a period of only five to twenty-five years.[4] That estimate ignores the growing world demand for power, as well as the Hubbert's peak effect, which is just as valid for uranium as for oil.

*The reactor that melted down in the Chernobyl power plant used graphite—that is, carbon—instead of water, though graphite reactors are generally used to make weapons-grade materials rather than electric power. However, there is a reactor design that uses graphite, tiny fuel particles, and helium gas cooling that is thought to be intrinsically safe against Chernobyl-type meltdowns. It's called a High Temperature Gas Reactor, or HTGR, and it can incorporate thorium as well as uranium in its fuel supply.

But that estimate assumes that only ^{235}U will be used to produce nuclear power. The vast majority of Earth's uranium is ^{238}U, which is not used at all in conventional power reactors. It is possible to design a reactor in such a way that while it is producing heat, it is also converting ^{238}U to the isotope 239 of the element plutonium. Like ^{235}U, ^{239}Pu is fissile; that is, if it captures a slow neutron, it can explode apart, yielding energy and neutrons. That kind of reactor is called a breeder reactor. Breeder reactors would increase the energy available from uranium a hundredfold. Unfortunately, plutonium is very nasty stuff; in particular, it is much more easily converted from peaceful pursuits into weapons than is uranium. For that reason, there are no commercial breeder reactors in the United States, but Russia, Japan and India have experimental plutonium breeder reactor programs. Making the world safe for breeder reactors is a tall order.

There is at least one other nuclear possibility. Isotope 232 of the element thorium can be bred into isotope 233 of uranium. ^{233}U is also fissile. Nuclear reactors using thorium as fuel are still rare, but thorium is thought to be an important nuclear resource for the future because it is about three times as abundant on Earth as uranium. The natural radioactive decay of thorium and uranium is what keeps Earth's interior hot.

Even if there is enough nuclear fission fuel on Earth to last for a while, the scale of what is needed is staggering. The largest practical nuclear power plant would produce about one Gigawatt (one billion watts) of power. Just to replace the 10 Terawatts of fossil fuel the world burns today would require opening 10,000 new Gigawatt plants, one a day for 30 years. And by then, of course, we would need to replace far more than 10 Terawatts.

NUCLEAR FUSION

The most promising, and most elusive, solution to all our problems is controlled nuclear fusion. The fuel it uses would virtually last forever and would not contribute carbon dioxide to the atmosphere. The easiest reaction to create is the fusing of two isotopes of the lightest element, hydrogen, called deuterium and tritium. (Only hydrogen isotopes have names of their own; all other isotopes go strictly by the numbers.) The nucleus of ordinary hydrogen is a single proton. The nucleus of deuterium is one proton with one neutron attached. The tritium nucleus, a proton with two neutrons, is rarely found on Earth, but it can be made: It is radioactive and lasts for only a few years before it decays spontaneously.

If a deuterium nucleus comes into contact with a tritium nucleus, the result is a helium nucleus (two protons, two neutrons) and an extra neutron. The extra neutron is fired off, releasing a great deal of energy. The problem is getting nuclei of the two isotopes together. Because both have positive electric charge, they repel each other with tremendous force, which increases as they get closer. But if they can be contained in a gas that is sufficiently dense and sufficiently hot, their random thermal motions can be fast enough to overcome the force of repulsion and they can be packed closely enough to have a good chance to fuse. We know it can happen. Nuclear fusion happens in the Sun.

In the Sun, the superheated gases that produce nuclear reactions are contained by the Sun's gravity. To fuse and create power on Earth, they must be contained in some other way, since no known material can stand the extreme temperatures that fusion generates. One solution to that problem is a magnetic field, which acts like a somewhat leaky virtual bottle, preventing the superheated, electrically charged particles from escaping, at least for a while. All of this takes place inside a vacuum chamber, keeping the hot stuff away from any material walls. So the trick is to make a gas of deuterium and tritium in a magnetic field—a gas hot enough, dense enough, and contained long enough to produce useful energy. But it takes a tremendous amount of energy to heat the gas to the point where that can happen and an enormous magnetic "bottle" to contain the gas long enough. Fusion reactions have been created this way in laboratories on Earth, but never enough of them to produce as much energy as it took to make the fusion reaction happen. That so-called break-even point has been the central goal of the field of fusion research for decades. Although remarkable progress has been made, the best that has been done on Earth so far is to get back about half the energy that went in.

In March of 1989, two scientists at the University of Utah announced what seemed to be the scientific discovery of the century. Nuclear fusion, producing measurable amounts of heat, could be induced to take place on a tabletop by electrolyzing deuterium-containing water using electrodes of platinum and palladium. This was known as cold fusion, and it seemed to be the solution to all our energy problems. After a five-week frenzy, the scientific community decided that cold fusion had all been a big mistake. A handful of scientists refused to go along with the consensus and continue to try to produce cold fusion to this very day, but they have not come up with a convincing demonstration that it actually works. Without scientific proof of its existence, we cannot know what technological importance cold fusion might have.

Generating usable hot fusion power will require a huge facility. A collaboration with Canada, the European Union, Japan, and Russia that might produce an engineering prototype of such a facility, called the International Thermonuclear Experimental Reactor, suffered a setback in 1998, when the United States withdrew because of the $10 billion price tag;

however the U.S. rejoined in February 2003, the ITER collaboration continues, and there is still hope that other countries will join it.

If magnetically contained nuclear fusion ever does become practical, the primary fuels will (at least at first) be deuterium and lithium. Tritium can be made in place from reactions of the outgoing neutrons with a lithium blanket around the reactor. Deuterium in abundant supply is found in seawater, and very likely we would never run out of it. Less is known about the supply of lithium, but lithium is found in a number of ordinary minerals.

Different approaches to containment are used by the National Ignition Facility at Lawrence Livermore National Laboratory and the Z facility at Sandia National Laboratory, in government-sponsored programs that are attempting to produce short bursts of fusion energy by heating deuterium and tritium fuel pellets with intense laser pulses (LLNL) or electrical discharges (Sandia). Like magnetic containment, these so-called inertial containment schemes are big and expensive. They, too, have their optimistic supporters, who hope they will reach break-even in the next decade. However, reaching break-even would still leave us a long way from a practical power plant.

SOLAR POWER

Beyond fossil fuel and nuclear power, all that remains is sunlight. We have, of course, always used sunlight—for example, to grow trees whose wood can be burned for warmth. One indirect solar source is hydroelectric power, a good example of seemingly renewable energy. The enormous pressure of water in a reservoir provides the force to drive a water turbine, which generates electricity. Then the Sun causes the water to evaporate, lifting it up into the clouds, whence it can rain into the watershed and flow into the reservoir to produce more electricity.

Early in the twentieth century, hydroelectric power seemed to be a great way to go, and dams to produce hydroelectricity were built wherever the appropriate conditions existed. Today about 10 percent of U.S. electric power and about a quarter of the world's electric power is generated that way. However, in recent years we have come to appreciate some of the disadvantages of damming up rivers, so that today, worldwide, dams are probably being disassembled faster than they are being built. Furthermore, we have virtually come to the end of our ability to increase the generation of hydropower. Dams have already been built pretty much wherever in the world they can be, so we cannot increase our production of hydropower enough to replace our dependence on fossil fuel. Finally, hydropower is not truly renewable, in the sense of a resource that will last forever. All reservoirs eventually silt up. After a few hundred years, the mighty Hoover Dam will be nothing but a concrete waterfall.[5]

Wind power is another indirect form of solar energy. About one quarter of one percent of U.S. electric power is now generated by the wind. That amount will increase, because technological improvements and tax breaks for power producers using renewable sources have made wind power economically competitive with coal-fired plants. In Northern Europe, where there's plenty of wind and fossil fuels are expensive, wind may someday rival hydropower as a source of electricity. However, many people regard wind farms as ugly and undesirable, and there are only a limited number of places in the world where the wind blows strongly and steadily enough to be useful. We may inherit the wind, but we won't be able to live on it.

Solar cells can convert sunlight to electricity directly. However, gathering solar energy at Earth's surface is a little like gleaning wheat from already harvested fields. The flux of light that reaches the surface of the planet is relatively weak and intermittent at best. Above the atmosphere, it is about eight times what it is, on average, at the surface. The loss of intensity is partly due to reflection and absorption of energy by clouds and the atmosphere, but mostly it is because the light gets spread out over the spherical surface of the spinning Earth. This observation has prompted the conception of a number of schemes that would intercept the Sun's light in space and transmit the energy to Earth.

One idea, studied in the 1970s under the sponsorship of NASA and the U.S. Department of Energy, was to put an array of solar cells about the size of Manhattan into a so-called geosynchronous orbit—that is, an orbit high enough so that the array would remain fixed above a spot on Earth. It would of course always face the Sun and would be positioned so as to stay out of Earth's shadow. Its electrical energy would be transmitted to Earth in the form of microwaves, a part of the electromagnetic spectrum used for radar because of its cloud-piercing properties. Proponents estimate that more than half the electric energy generated at the solar array could be converted into electric energy in a receiving station on Earth roughly 8 miles by 6 miles in area. About eight hundred such satellites in orbit, each with a receiving area on Earth, would be needed to provide as much energy as we use on Earth today.[6] Other schemes of this sort have been studied more recently, using more advanced technology. It is conceivable that space-based solar power may someday make some contribution to world energy resources.

IMPROVING WHAT WE'VE GOT

The best, most conservative bet for ameliorating the coming fuel crisis is the gradual improvement of existing technologies. To take just one modest example, at the end of the nineteenth century Thomas Edison's invention of the incandescent light-bulb became the very symbol of the dawning age of electricity. The incandescent lightbulb works by passing an electric current through a thin resistive filament, heating it to a temperature so high that it radiates white light, like the face of the Sun. That is to say, the principal product of a lightbulb is not light but heat. Producing heat is the most wasteful possible way of using electric energy. Only 1 or 2 percent of the electric energy consumed by an incandescent bulb turns into visible light.

A variety of more efficient light sources have been invented since Edison's day, although incandescent bulbs still rule the night. One new type—the light-emitting diode (LED)—is worth a careful look. In certain kinds of materials called semiconductors (the stuff of transistors), an electric current doesn't just generate heat; instead it causes quantum mechanical events in which individual electrons absorb a well-defined quantity of energy from the current. When the electrons fall back into their original states, they give off, in the form of photons, all the energy they absorbed. Photons are light, and depending on the particular properties of the semiconducting material, they can be visible light of various colors.

A few years ago, LEDs were so inefficient that their only application was in alpha-numeric displays, like the flashing "12:00" on the face of your VCR that shows that you haven't bothered to learn how to program the clock. The problem was not in the initial *electric current → excited electron → visible photon* reaction, which works very well, but rather in getting the photon out of the material before it bounced around many times and turned into heat. But clever engineers have solved that problem. As you drive around town, you may notice that the old-fashioned incandescent traffic lights are gradually being replaced by much brighter ones that have a kind of speckled look. Those are arrays of LEDs, which are so efficient that the savings in electric power easily makes up for the cost of buying the expensive devices.

Solar cells—also called photovoltaic devices, or PVs—are just LEDs running backward. In an LED, you put electric current into a semiconducting device and light comes out. In a PV, you put light into the same kind of semiconducting device and electric current comes out. That would suggest that PVs can now be made very efficient, turning all the light that falls on them into electric current, but this isn't the case. The reason is that the LEDs are efficient only at a single color—red traffic light, green traffic light, and (less common at present) amber traffic light—depending on the particular semiconducting material they are made of. A PV, to be efficient, must be able to turn all of the light falling on it, of all possible colors, into electrical energy. PVs that can do that do exist, but they are very expensive and are currently used only in spaceflight applications.

The lesson to be learned from this story is that the most exotic objects (LEDs) can become commonplace (traffic lights) almost without our noticing it, and they can suggest the direction of future developments (cheap, efficient PVs). Such a development could, for example, make the Saudi Arabian desert more valuable for the sunlight falling on it than for the oil buried beneath it. However, the scale of what is needed is breathtaking. Using present-day PV technology, in order to replace all the power generated from fossil fuels, an array spread over more than 200,000 square kilometers would be needed. That's an area roughly half the size of the State of California. All the PVs made up to now would probably cover fewer than 10 square kilometers.[7]

In 1986, the scientific world was astonished to learn of the completely unexpected discovery of high-temperature super-conductivity. An earlier scientific world, around 1911, had been even more astonished to learn of the discovery of the phenomenon of low-temperature superconductivity. It turned out that at temperatures a few degrees above absolute zero, many metals quite suddenly become capable of conducting electricity with no resistance at all. The 1986 discovery was that certain complex materials were capable of performing the same trick at a much higher temperature (albeit one still very far below ambient temperature anywhere on Earth).

The discovery of high-temperature superconductivity immediately evoked pictures of a worldwide electric energy grid, which would serve to balance demand for electric power between day and night. As noted in chapter 3, electric energy is hard to store in large quantities and so must be generated on demand, which is much higher during the day than at night. Since it is always day on one part of the planet and night on another, a worldwide electric grid would solve that problem. However, electric power lines using high-temperature superconductors have not yet proved feasible. They are subject to the same drawbacks as were earlier plans to use low-temperature superconductors for power transmission (the need for elaborate refrigeration and insulation, back-up systems in case of catastrophic failure, and so on). To make matters much worse, high-temperature superconductors are composed of more exotic, more expensive materials than the old, low-temperature ones. Still, the astonishing discovery of high-temperature superconductivity serves to remind us that nature may still have surprises for us that can change the landscape.

As this brief survey suggests, there is no single magic bullet that will solve all our energy problems. There is no existing technology capable of replacing the oil we will soon be without, nor is there any on the horizon that we can depend on to replace the remaining fossil fuels when they are exhausted. And if we permit them to become exhausted before replacing them, we may place the climate of our planet in grave danger. The best hope for our civilization lies in technologies that have not yet arisen—possibly based on scientific discoveries that have not yet been made. Most likely, progress will lie in incremental advances on many simultaneous fronts, based on principles we already understand: controlled nuclear fusion, safe breeder reactors, better materials for manipulating electricity, more efficient fuel cells, better means of generating hydrogen, and so on. Developing those technologies will require a massive, focused commitment to scientific and technological research. That is a commitment we have not yet made. We urgently need to make it. ⊕

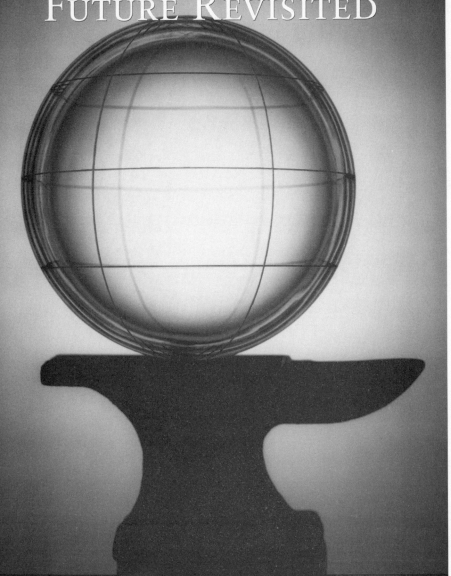

ENVOY: THE
FUTURE REVISITED

Four and a half billion years ago, chunks of rock that coalesced out of the primordial dust cloud circling the Sun clumped together to form Earth, a spinning sphere in a nearly circular orbit 93 million miles from the Sun. Those facts alone did not ordain our planet's destiny. It could have become nothing more than an icy wasteland, reflecting back into empty space most of the light from the Sun that falls on it—and, indeed, at some periods in its deep past it may have been just that. It could have turned into a poisonous, lifeless inferno, like its near twin, Venus. Instead it has become a balmy garden planet, with an oxygen-rich atmosphere and much of its carbon sequestered in the ground in the form of coal and other fossil fuels. It was life itself that helped to create this planetwide Garden of Eden.

Life on Earth is sustained by a great underlying drama of radiant energy from the Sun—radiant energy that drives vast currents in the atmosphere and oceans, warms the soil, and in a thousand other ways undergoes its inexorable entropic transformation into ambient-temperature heat, eventually to be radiated back out into space as invisible infrared rays. But that exchange of radiant energy happens on Venus, Mars, and every other orbiting body in our solar system without nourishing life along the way. Now, in the last few instants of those four and a half billion years, life has turned intelligent enough to dig up and burn those fossil fuels. Humans will not be banished from the Garden of Eden, but we may very well wind up destroying it.

In the meantime, we have a much more pressing and immediate problem. Our way of life, firmly rooted in the myth of an endless supply of cheap oil, is about to come to an end. But is it possible that this view is mistaken? Are there not experts on the other side who believe that the end of the age of oil is still decades away?

Yes, there are. We have already discussed the 1995–2000 USGS study. The oil companies do studies of their own. BP is one of the world's more forward-looking major oil companies. It publishes a Web site full of useful information, not only on oil but also on other fossil fuels, nuclear energy, and renewable sources of energy.[1] According to the BP Web site, the ratio of known reserves of oil to rate of use—the R/P ratio described in chapter 1—is forty years. The R/P ratio for natural gas is sixty years.

The U.S. Department of Energy, citing the USGS study, gives the R/P ratio for oil as ninety-six years.[2] The United States does not have a Department of Entropy.

The R/P ratio cited by BP is based on data easily found on their Web site. In the figure below, the pie chart on the left shows where the "proved oil reserves" are to be found. The vast

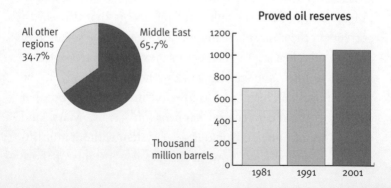

majority, nearly two thirds, are in the Middle East. No surprise there. The bar chart on the right gives the known oil reserves as of 1981, 1991, and 2001, in thousands of millions (i.e., billions) of barrels of oil. The amount in 2001 was just over one trillion barrels. In other words, as of 2001, more than half the two trillion barrels of oil that ever existed was sitting in the ground in known reserves, just waiting to be pumped. We're currently consuming about one-fortieth of that amount per year. Therefore as of 2001 we seemed to have a forty-year supply. How did the Hubbert's peak gang come to be so deluded? Are they using different, less reliable data?

They are using exactly the same data, and to the practiced eye the justification of their predictions can be seen directly on the BP graph. The known reserves increased from 0.7 trillion barrels in 1981 to 1 trillion barrels in 1991 and to about 1.03 trillion in 2001. This slowing down of the rate of increase is a consequence of just the phenomena that led to the Hubbert's peak predictions. In fact, even the increase after 1981 may have been largely due not to any real discovery of new oil but to a change in OPEC rules that caused its member countries to increase their reported reserves in the 1980s.[3] The rate at which oil is discovered is a bell-shaped curve that peaked decades ago; the rate at which oil can be pumped out of the ground is the bell-shaped curve whose peak the Hubbert followers are trying to predict. The quantity shown in the BP graph is the amount known to be still in the ground. It is the difference between how much has been discovered and how much has already been pumped out. When the rate of increase of known reserves reaches zero (which for all practical purposes may already have happened), we will for the first time in history be consuming oil faster than we are finding it.

The crucial difference between the Hubbert's peak prediction and the BP prediction is not a matter of how much oil there is, or of how fast we're using it. They pretty much agree on those questions. The difference is in when the crisis will occur. The unstated assumption of the BP prediction is that we'll be fine until the last drop of oil is pumped out of the ground. The Hubbert's peak prediction says that once we reach the halfway point in consuming the oil in the ground, existing oil fields will start to become exhausted faster than new oil fields can be tapped. The rate at which oil can be pumped out of the ground will start to decline. That is the essence of the bell-shaped-curve hypothesis. The Hubbert's peak assumption is that the crisis will occur not when the last drop is pumped but at the halfway point, where falling supply meets continuing rising demand. If we have already consumed nearly half the oil there ever was, the crisis can't be far off.

There is no doubt at all that the essence of the Hubbert peak view is correct. It is possible, of course, that the quantitative predictions are off, so that the crisis won't occur until the next decade or even the one after that. That difference might seem important to us, but in the long view of history a difference of ten or twenty years means nothing at all. We, or our children, or our grandchildren face some very difficult times.

If the problem were widely understood and acknowledged, we could go a long way toward easing the pain that the crisis will cause. We Americans are profligate users of energy. There are many ways in which we could reduce our consumption of fuel without abandoning our comfortable way of life. That would give us more time to convert to a temporary methane-based technology, while we build up our capacity for tapping other fuel supplies.

In the long run, even those steps will not be enough. The real challenge—the challenge we would set for ourselves if we had courageous, visionary leadership—would be to kick the fossil fuel habit altogether as soon as possible. In 1960, John F. Kennedy challenged us to put a human being on the moon within that decade. And we did it! That was possible because we already knew the basic principles of how it could be done. There were formidable technological obstacles to overcome, but we are very, very good at overcoming that kind of obstacle when we put our minds to it. The energy problem is of exactly that nature.

We can envision a future in which we live entirely on nuclear energy and solar energy as it arrives from the Sun. That would not require a reversion to an eighteenth-century lifestyle and a concomitant drastic reduction of the human population of the Earth. Instead it would be based on a sophisticated technology that converts sunlight and nuclear energy efficiently into electricity for stationary uses, and produces hydrogen fuel or charges advanced batteries for mobile uses. That would leave the carbon in the ground, or at least unburned, as a source for the petrochemicals that are also an indispensable feature of our way of life. And it might be no more difficult to accomplish than putting people on the moon.

Unfortunately, our present national and international leadership is reluctant even to acknowledge that there is a problem. The crisis will occur, and it will be painful. The best we can realistically hope for is that when it happens, it will serve as a wake-up call and will not so badly undermine our strength that we will be unable to take the giant steps that are needed.

For now I stand by the warning I made in the first paragraph of this book's Introduction: Civilization as we know it will come to an end sometime in this century unless we can find a way to live without fossil fuels.

Postscript

enneth Deffeyes, a leading academic geologist, well schooled in the realities of oil and gas, realized that geophysicist M. King Hubbert had been right—that is, a peak in oil production for the Lower 48 had been reached—when he read a brief sentence in the *San Francisco Chronicle* in the spring of 1971: "The Texas Railroad Commission announced a 100% allowable for next month."[1] Such a quiet pronouncement would have slipped by most readers, but to an insider like Deffeyes the words were momentous. The quaintly named Texas Railroad Commission, after all, was the cartel that controlled production in the U.S. oil industry by manipulating the excess pumping capacity of Texas wells. By announcing a "100% allowable," the commission signaled that Texas no longer had any excess pumping capacity. The Texas oil fields could be pumped flat out because they had reached a point of diminishing return—and consequently the Texas Railroad Commission had lost control of the market.

After that, the world oil market fell under the domination of another cartel, the eleven-nation Organization of Petroleum Exporting Countries, indeed modeled after the Texas Railroad Commission, and led by Saudi Arabia. The Saudis have been manipulating the price of oil ever since by strategic use of their excess pumping capacity. Thus the news to look for, signaling the arrival of a *worldwide* Hubbert's peak, would be that Saudi Arabia no longer had any excess capacity.

Exactly that news appeared in a story on the front page of the *New York Times* on February 24, 2004. Headlined "Forecast of Rising Oil Demand Challenges Tired Saudi Fields" and written by Jeff Gerth, the story went on to say, "The country's oil fields now are in decline, prompting industry and government officials to raise serious questions about whether the kingdom will be able to satisfy the world's thirst for oil in coming years." Like many *New York Times* articles, this one was very long and regularly contradicted itself in an apparent effort to achieve what's called balanced reporting. So, much farther along, Gerth wrote, "Some economists are . . . optimistic that if oil prices rise high enough, advanced recovery techniques will be applied, averting supply problems." In the very next paragraph Gerth takes it back: "But privately, some Saudi oil officials are less sanguine." And so on. It will be a long time before we'll be able to say whether February 24, 2004, was indeed the beginning of the end of the age of oil, but to those familiar with the Hubbert's peak predictions (as Gerth apparently was not), it was a chilling story.

OPEC's policy over recent decades has been to control oil supply so as to keep the price of oil not just above some chosen minimum but also within a certain range—not too low but also not too high. At this writing, the stated desired range is $22 to $28 a barrel but, since the current price is considerably higher, OPEC is said to be considering upping the range. The reason for trying to keep the price from going too high is partly in order not to discourage demand for oil, but also to prevent investment in alternative fuels. The implied threat is, if you invest money to develop a competitor to oil, we will flood the market with cheap oil and wipe out your investment. However, if the Saudi fields have really peaked, that becomes an empty threat, and the cartel stands to lose control of the market.

The United States is not a member of OPEC, but its government shares OPEC's goal of keeping an upper limit on the price of oil; voters tend to get very unhappy when the price of gasoline at the pump rises. If Saudi Arabia can no longer flood the market, where would the extra oil supply be found? Canada (also not an OPEC member) now claims the world's second largest reserves after Saudi Arabia, but those are largely locked up in oil sands, solid deposits that must be mined, not pumped, and so will not be flooding anything anytime soon. The world's third largest reserves are claimed by Iraq, a hundred billion barrels of liquid oil waiting to be exploited. Under the regime of Saddam Hussein, however, the spigot was broken. Although nobody seems willing to talk about it, that was certainly one of the big reasons for the U.S. invasion of Iraq in 2003: The idea was not so much to steal oil from the Iraqi people—they will be allowed their small profit for the raw material before the real money gets made. Rather, the idea was simply to get Iraqi oil back on the market.

Some experts doubt that there will be an oil crisis in the near future. They have been dubbed the "antidepletionists."[2] In my experience they are intelligent, well-informed people—and most of them are employed by the oil industry. That doesn't automatically make them wrong. After all, people who work in the oil industry are the ones most likely to be interested and knowledgeable about it. We should keep in mind, though, that the oil industry has a very strong incentive to deny any looming shortage of oil. The reason is to keep down the price of oil properties they would like to acquire.

As we have seen, the worldwide "proven reserves" of oil now stand at just over one trillion barrels, and the R/P (reserves-to-production) ratio is about forty years. Nothing alarming about that, say the antidepletionists; the R/P ratio hovered around forty years through most of the twentieth century. That is true, but to understand what the ratio really means we have to reexamine the term "proven reserves." To most of us, "proven reserves" would consist of all the oil that's been discovered minus all the oil that's already been extracted. But that is not how the oil industry uses the term. Oil companies and petroleum-producing nations alike report as "proven reserves" only a portion of what they believe themselves to have in reserve. When a new field is discovered, geologists use various techniques to measure its length and width, its depth, the porosity of the rock, and so on, finally coming up with an estimate of how much oil the field might contain. That estimate gets turned over to officials of the company or country, who can report as "proven" whatever fits their current needs, saving the rest for a rainy day. That leeway is what permits "proven reserves" to go on growing and the R/P ratio to remain essentially constant no matter what is happening in real oil fields.

What is actually going on in real oil fields is sobering. Worldwide, the rate of discovery peaked around 1960 and has been declining ever since. Meanwhile, the worldwide rate of consumption of oil has continued to grow, first exceeding the rate of new discovery around 1980. The gap between the two has grown steadily during the last twenty-five years. That should mean that proven reserves have declined by some hundred and fifty billion of barrels over that period. Instead, the reserves have steadily increased. Why? Because companies and countries continue to pull out new reserves that they've kept up their sleeves. In fact, in the late 1980s the proven reserves of OPEC nations jumped by nearly four hundred billion barrels without the benefit of *any* new discoveries. To reach that new height, OPEC merely changed its quota rules for how much oil each member nation was permitted to pump based in part on their reported proven reserves, and the new proven reserves magically appeared.

Add the growing gap between rate of discovery and rate of consumption to the giant jump in OPEC reserves and we see that something like five hundred billion barrels of oil have been brought out of the shadows by these methods and added to worldwide proven reserves over the past twenty-five years—an amount equal to roughly half of all existing reserves. Obviously this game can't go on much longer. Either the industry will run out of hidden reserves or they will simply start lying—reporting reserves that don't exist. That may have already started to happen. The once proud Royal Dutch Shell Group recently made headlines when it was forced by outside auditors to reduce its claims of proven reserves—and correspondingly the value of its stock shares.

Antidepletionists are fond of saying that discovery has been declining since 1960 because so much oil had already been found that no more was needed; thus exploration dwindled to a standstill. That is most certainly not the case. For example, 1999 and 2000 were spectacular years for oil discovery, driven by giant findings at Azedegan in Iran and the Kashagan East field in the North Caspian Sea.[3] But even in those years, new discovery fell far short of consumption. In truth, the world is consuming oil at such a breathtaking rate—more than twenty-five billion barrels per year and rising rapidly—that no discoveries, past, present, or future, are going to keep up with demand. And remember, the people of China are just beginning to drive.

Economists believe that the demand for anything can never exceed its supply. The mechanism of price assures that the supply will show up when it's needed. Of course, that has pretty much never been true of the oil industry, which has nearly always been governed by cartels, first the Texas

Railroad Commission, then OPEC. When world oil production peaks, OPEC will lose control, and the price mechanism will kick in with a vengeance, making it economically feasible for other sources of fuel to replace the missing oil. In a sense, that has already happened in the case of Canadian oil sands, which are now being mined at a profit. But the product that comes out of the ore is not rich enough to make gasoline, so hydrogen must be added. As a result, some of the world's largest plants for extracting hydrogen from natural gas have been built in Alberta. In other words, oil from oil sands is not only more costly in money than conventional oil, it is also more costly in energy. That will be increasingly true as other hydrocarbon resources are exploited.

Thus, if we are willing to let the planet's climate fend for itself while we go merrily burning fossil fuels at ever-increasing rates, and if we are willing to pay ever-higher prices in both cash and energy, we may be able to muddle through for much of the coming century—that is, provided global political and social stability can somehow be maintained in the face of huge fuel costs and the economic dislocations that will entail.

Meanwhile, in the current (possibly) pre-peak period, our enormous consumption of conventional oil makes us wholly dependent on some pretty dicey places in the Middle East. Perhaps truly at issue here is not the debate between depletionists and antidepletionists but quite a different kind of question: Which comes first, Hubbert's peak or the collapse of the Saudi regime? Both would have the same effect, and both seem inevitable.

To sum up, there are at least three good reasons for trying to kick the fossil fuel habit as soon as possible: First, our present dependence on cheap oil makes us subject to events in some very unstable parts of the world; second, burning up all the fossil fuels we can get our hands on could cause irreversible damage to the climate of the only planet we have; and third, since the stuff will eventually run out in any case, we should give ourselves the best head start we can in preserving all that is worth preserving in our civilization. Kicking the fossil fuel habit will require harnessing and organizing the creativity and ingenuity of scientists, engineers, social scientists—indeed, all of us—all over the world.

This is no small task. We do understand the underlying scientific principles that will solve the problems, but we don't know what kinds of solutions will prove both technically possible and socially feasible. We can't let the choice of solutions be dictated by some central authority; neither can we afford to leave a solution up to the marketplace. The challenge is enormous but the stakes are even larger. If future generations are to thrive, we who have consumed Earth's legacy of cheap oil must now provide for a world without it.

NOTES

INTRODUCTION

1. See Kenneth S. Deffeyes, *Hubbert's Peak: The Impending World Oil Shortage* (Princeton N.J.: Princeton University Press, 2001), p. 158; A. A. Bartlett, "Reflections on Sustainability, Population Growth and the Environment," *Population & Environment* 16, no. 1 (September 1994), pp. 5–35; Colin J. Campbell and Jean H. Laherrère, "The End of Cheap Oil," *Scientific American,* March 1998; Richard C. Duncan, "World Energy Production, Population Growth, and the Road to the Olduvai Gorge," *Population & Environment* 22:5 (May–June 2001); L. F. Ivanhoe, www.hubbertpeak.com/ivanhoe.

CHAPTER 1: THE FUTURE

1. See Deffeyes, *Hubbert's Peak,* p. 1; "Hubbert, Marion King," The Handbook of Texas Online. www.tsha.utexas.edu/handbook/online/articles/view/HH/fhu85.html, accessed Feb 4 2003.

2. Deffeyes, *Hubbert's Peak*, p. 157.

3. This point is made by Michael Lynch, chief energy economist, DRI-WEFA, Inc., at http://sepwww.stanford.edu/sep/jon/world-oil.dir/lynch/worldoil.html and http://sepwww.stanford.edu/sep/jon/world-oil.dir/lynch2.html.

4. U.S. Department of Energy predictions of this kind can be found at http://www.eia.doe.gov/pub/oil_gas/petroleum/presentations/2000/long_term_supply/sld001.htm.

5. For further examples see Walter Youngquist, *Geodestinies* (Portland, Ore.: National Book Company, 1997).

6. This point was made by Alec N. Brooks in a paper titled "Perspectives on Fuel Cells and Battery Electric Vehicles" presented at the CARB ZEV Workshop, December 5, 2002. CARB and ZEV stand for California Air Resources Board and Zero-Emission Vehicles.

CHAPTER 2: ENERGY MYTHS AND A BRIEF HISTORY OF ENERGY

1. See www.cmhrc.pwp.blueyonder.co.uk, accessed December 17, 2002.

2. See www.chernobyl.co.uk, accessed January 19, 2003.

3. An excellent description of Joule's classic experiment is his letter to the editors of *Philosophical Magazine*, which is found at http://dbhs.wvusd.k12.ca.us/Chem-History/Joule-Heat-1845.html.

CHAPTER 3: ELECTRICITY AND RADIANT ENERGY

1. Quoted in H. W. Brands, *The First American* (New York: Anchor, 2000), p. 205.

2. S. Arrhenius, "On the Influence of Carbonic Acid in the Air upon the Temperature of the Ground," *Philosophical Magazine and Journal of Science*, April 1896.

3. Martin I. Hoffert et al., "Advanced Technology Paths to Global Climate Stability: Energy for a Greenhouse Planet," *Science* 298 (Nov. 1, 2002): 981–987.

4. John Wettlaufer, private communication.

CHAPTER 4: HEAT ENGINES AND ENTROPY

1. Rudolf Clausius, "Ueber verschiedene für die Anwendung bequeme Formen der Hauptgleichungen der mechanischen Wärmetheorie," *Annalen der Physik und Chemie*, 125 (1865): 353–400, translated and excerpted in William Francis Magie, *A Source Book in Physics* (New York: McGraw-Hill, 1935).

CHAPTER 5: TECHNOLOGICAL FIXES

1. This suggestion, together with many other ideas discussed in this chapter, can be found in Martin I. Hoffert et al., "Advanced Technology Paths."

2. See R. A. Metzger, G. Benford, and M. I. Hoffert, "To Bury or to Burn: Optimum Use of Crop Residues to Reduce Atmospheric CO_2," *Climatic Change* 54 (2002): 369–374; and "Sequestering of Atmospheric Carbon Through Permanent Disposal of Crop Residue," *Climatic Change*, 49 (2002): 11–19.

3. www.planetark.org/dailynewsstory.cfm/newsid/17543/ newsDate/ 2-Sep-2002/story.htm, accessed February 25, 2003.

4. See Hoffert et al., "Advanced Technology Paths," p. 985.

5. Walter Youngquist, "Alternative Energy Sources," in L. C. Gerhard, *Sustainability of Energy and Water Through the 21st Century* (Lawrence, Kan.: Kansas Geological Society, 2002).

6. See Hoffert et al., "Advanced Technology Paths," p. 985.

7. Ibid.

ENVOY: THE FUTURE REVISITED

1. www.bp.com/centres/energy2002.

2. E-mail reply to a query by the author to a question-answering service run by the Department of Energy. The e-mail address of the service is Infoctr@eia.doe.gov.

3. See Deffeyes, *Hubbert's Peak,* p. 147.

POSTSCRIPT

1. See Deffeyes, *Hubbert's Peak*, p. 4.

2. Bob Williams, "Timing Is Everything: The Information Crisis on Future Oil Supply," Offshore Technology Conference, 2004.

3. Zagar et al., "The Iron Grip of Depletion," Offshore Technology Conference, 2004

ANNOTATED BIBLIOGRAPHY

The scientific background of much of this book can be found in *The Mechanical Universe and Beyond*, a series of fifty-two half-hour television programs that can still occasionally be seen on PBS stations in the middle of the night, for which I was creator and host.

At a slightly more advanced level, the enduring classic of introductory physics is the three-volume *Feynman Lectures on Physics*, ed. R. P. Feynman, R. B. Leighton, and M. Sands, 25th anniversary commemorative edition (San Francisco: Addison-Wesley/Benjamin Cummings, 1989).

These days the World Wide Web serves as an almost infinite source of both information and misinformation. (Let the reader beware.) Just a few relevant sites:

The home page of the U.S. Geological Survey: www.usgs.gov/index.html;

The home page of the U.S. Energy Information Administration: www.eia.doe.gov/;

Hubbert's Peak: www.hubbertpeak.com/.

H. W. Brands, *The First American* (New York: Anchor, 2000). A good biography of Benjamin Franklin.

Kenneth S. Deffeyes, *Hubbert's Peak: The Impending World Oil Shortage* (Princeton N.J.: Princeton University Press, 2001). Deffeyes was an associate of M. King Hubbert when both worked for the Shell Oil Company in Houston. He is professor emeritus of geosciences at Princeton University, and this knowledgeable book is full of useful information and amusing stories about oil geologists and the oil industry.

Martin and Inge Goldstein, *The Refrigerator and the Universe: Understanding the Laws of Energy* (Cambridge Mass.: Harvard University Press, 1993), expands on a number of the themes in this book, but you'll have to put up with a few equations.

David Goodstein, *States of Matter* (New York: Dover, 2002). A graduate-level textbook that you certainly don't want to read unless you are an expert.

John L. Heilbron, *Electricity in the 17th and 18th Centuries* (Berkeley Ca.: University of California Press, 1979). The best reference for the early history of electricity.

Jeremy Rifkin, *The Hydrogen Economy* (New York: Putnam, 2002). A polemicist with a somewhat anti-science tinge, Rifkin does a good job of presenting the Hubbert problem, though his solution to it—that we will each generate our own hydrogen fuel using only renewable energy sources—is utopian, to put it kindly.

Hans Christian von Baeyer, *Maxwell's Demon* (New York: Random House, 1998). Attempts to write about thermodynamics at the popular level are rare. The Goldstein & Goldstein book is one, and this is another.

L. Pearce Williams, *Michael Faraday: A Biography* (New York: Basic Books, 1965). A classic biography and an excellent account of electricity in an era later than the Heilbron book.

Walter Youngquist, *Geodestinies* (Portland Ore.: National Book Company, 1997). A compendium of information about all kinds of natural resources, presented in a rather dry, straightforward way by an eminent petroleum geologist who has served as a consultant to a number of oil companies.

Mark Zymansky, *Heat and Thermodynamics* (New York: McGraw-Hill, 1937). Every undergraduate thermodynamic textbook cycles through the various heat engine cycles. Zymansky's is my own favorite, since I suffered it in college. I used, and still consult, the fifth edition, published in 1957.

ACKNOWLEDGMENTS

A number of colleagues and friends have read parts or all of this manuscript and made suggestions and comments that influenced subsequent drafts. Among those, Judith Goodstein, Brad Haugaard, and Robert Mackin deserve special thanks. I am also grateful for the contributions of Ferdinand Banks, Mark Goodstein, Roy Gould, Pierre Jungels, Steven Koonin, Joseph Krieger, Nathan Lewis, Paul MacCready, George Maltezos, James Morgan, Lynn Orr, David Stevenson, Thomas Tombrello, and John Wettlaufer. It has been a great pleasure to work with Nan Oshin and Jeffrey Atherton, who designed and illustrated the book. The book was edited with skill and patience by Sara Lippincott. Ed Barber at W. W. Norton provided essential encouragement.

Needless to say, any remaining errors are my responsibility alone.

INDEX